I0095259

The Architecture of Forever

Governing Life After Aging

Amos Behana

Singularity
PUBLISHING

Copyright ©2025 by Singularity Publishing LLC

All rights reserved.

This book is for informational purposes only. The author and publisher do not provide legal, financial, medical, psychological, or other professional advice. Readers should consult qualified professionals before making decisions affecting their health, finances, business, or personal matters. The author and publisher disclaim all liability arising from the use of this book.

For information, contact:

Singularity Publishing LLC

www.SingularityPublishing.com

ISBN: 978-1-969989-00-1 (Ebook) | 978-1-969989-03-2 (Paperback) | 978-1-969989-06-3 (Hardback) | 978-1-969989-09-4 (Audio)

Library of Congress Control Number (LCCN): 2025923224

Printed in the United States of America

First Edition

10 9 8 7 6 5 4 3 2 1

Contents

Author's Note

On the Name *The Continuity Sequence*

This book is the first volume in a larger work I have chosen to call *The Continuity Sequence*. I chose this name deliberately, not to describe a single technology or imagined event, but to name the problem that sits beneath all of them.

Much of our public conversation about the future is organized around moments of rupture, breakthroughs, revolutions, or thresholds after which everything is assumed to change. While these moments matter, they often distract from the larger, more difficult question of what comes next. If human lives can be extended, if machines can act with increasing autonomy, and if the systems we build are designed to persist, then continuity becomes the defining condition of the future. We are no longer preparing for endings as much as we are planning for endurance.

The word *sequence* reflects the structure of this work. Each book addresses a necessary stage in the unfolding of a problem. This volume examines continuity at the level of human life and civilization. The next turns to the delegation of agency to non-human systems. The final considers what stewardship, responsibility, and governance mean when humans are no longer the sole or even primary actors. These are not separate questions; they are steps in a single progression.

I have chosen *continuity* because it carries moral weight. To extend life, agency, or power without extending responsibility is to create systems that outlast our willingness to care for them. Continuity demands foresight, restraint, and stewardship. It asks not only what we can build, but what we are prepared to maintain.

This book stands on its own, but it is also an opening move. It is an attempt to think clearly about what endures once the limits we once took for granted begin to dissolve and about the obligations that emerge when the future stops ending where it used to.

Introduction: The Last Generation to Die

Sarah Taylor was born in 1995, the year the web entered most homes. She learned to code before she could drive, and she joined a robotics club with 3D printers older than herself. She graduated into a world with machine learning next to lab notebooks and clinical dashboards. At 30, she splits her time between a wet lab, where she grows tissue, and virtual rooms, where colleagues in Lagos, Bangalore, and São Paulo co-design protocols in real time.

Despite her outstanding credentials, what makes Sarah most remarkable is that she belongs to the first cohort for whom death from aging could become rare and even elective within their lifetime. Advances in genetic engineering, such as CRISPR technology, and breakthroughs in senescence research suggest a future where we have unprecedented control over biological aging, potentially making death from aging an unusual occurrence.

This book gathers evidence and clear markers to demonstrate that progress in three interconnected fields—artificial intelligence (AI), biotechnology, and neurotechnology—is making this shift possible. I don't refer to this outcome as a "singularity"; instead, I consider it a powerful convergence.

AI speeds up discovery and design. This progress is enabled by biotech that repairs and modifies bodies, bridging the gap between biology and machines. Meanwhile, neurotechnology can read and interpret signals sufficiently to restore function and, one day, support digital continuation. This means that it will be possible to ensure the preservation and extension of cognitive functions and identities through digital means, allowing aspects of human consciousness to be supported or augmented. Of course, this should be done for specific purposes under strict controls.

None of these advances is magic. Together, and with the proper oversight, they change mortality tables, clinical guidelines, curricula, and law.

This shift is not just made up of slogans; it's becoming visible in Sarah's world. In clinics, longer-lasting health care is becoming standard, not a special option. For this reason, people and institutions are paying for treatments that use a person's own cells. Specialists and physicians are using treatments, proven safe over several years, that slow down aging to treat certain health problems. Also, medicine is utilizing gene editing to lower the risk for rare diseases and for common problems, with ongoing monitoring. Tiny sensors can keep track of health data continuously to catch issues early. Public health records show that these changes are leading to fewer strokes, broken bones, and hospital stays over time.

These medical advances are not the only ones. Neural interfaces are devices that help people with injuries or illnesses communicate and control things. These devices can work well for many years. They are managed and checked in a similar way to other important medical devices.

Amid this optimism, I am aware of the limits of these technologies and need to express them explicitly. I do not consider "uploading" as solved; instead, I use the narrower term "digital continuation," and I insist on verification before taking any action on a claim. This book is not a promise of immortality; rather, I focus on what I term "elective mortality." Elective mortality refers to the ability to choose when and how one might encounter death by managing and delaying aging-related decline as a process that's separate from chronological age.

Connectomics is the study of neural connections in the brain at various circuit modeling scales. It carefully uses rich recordings from living brains to constrain architectures. It is assumed that the most sensitive area, digital continuation, has consequences; it isn't a metaphysical truth. A continuity agent is an entity or system tasked with ensuring ongoing identity in a digital form. It must pass preregistered and go through independently audited tests that go beyond style, including how the agent behaves under new circumstances. Its behavior must show consistency to be safe. Other tests include recalling personal memories with the right details and feelings, checking if traits and values stay the same over time, confirming facts with people who know the person well, and clear tracking of information from collection to use. Even then, recognition works only in certain situations, and permission can be withdrawn when possible or required.

The law and other institutions are expressly adapting statutes and public dashboards. Progressively, more legal systems are safeguarding end-of-life choices, which are being legalized in more jurisdictions. For example, consider a new law in a midsize city that updates estate planning practices. Wills include digital assets, like online accounts and digital profiles, and the ability for secure identity systems to access their management. The new legal framework updates laws about estates and guardianship to include digital profiles while protecting the rights of those who are still alive. Identity registries anchor obligations and rights to specific instances instead of just using a name, which may cover many individuals. Actuarial tables move. Courts now consider digital identity during trials, ensuring a comprehensive representation of a person's legacy. Other innovations include reimbursement codes for regenerative protocols, while medical curricula cover maintenance-of-practice for careers spanning a century, complete with gaps and pivots.

In this scenario, equity must be part of the design from the beginning, not added later. Access to and outcomes of longevity care, neural interventions, and continuation services will vary by region and income level, and ambitions of immortality will ultimately depend on real infrastructure and budgets. Then, energy and reliability will become ethical variables with regard to hospitals, water systems, and data centers, which must procure clean power, run drills, and provide public reports after actions have been taken in words the public can understand.

If these markers seem procedural, that's intentional. This book is a work plan rather than a wish list. It sets clear guidelines to treat optimism about the future seriously. In the chapters ahead, you will see the same practices throughout:

- preregistered trials and independent audits

- long-term registries linking biomarkers to real outcomes

- consent as an ongoing process

- clear tracking from data to model to deployment

- the rule of "one identity means one civic vote," even as lives multiply

- open access and transparency to keep digital life open

- stewardship dashboards to ensure that longer lives mean more care

- regular drills to prove reliability

The usual way of dying has become less common, offering individuals more control over their lifespan without seeking eternal life. Our focus is on agency, not eternity. I also resist determinism. The true questions are "Where?" "How fast?" and "Under what rails?", and whether a single moment alters the course of the future from what it was to what it will become.

The book begins at a human scale, sharing real-life scenes from people who could be you or me. You'll meet Maria, who wonders if her brother's funeral was the family's last natural death. You'll read about Aminatou, from a city where candles are lit for lives cut short by a lack of resources. Lena brings a copy home to dinner and finds that love, law, and ledgers must all adjust.

Throughout these pages, you'll also see checklists and vignettes, markers to track reimbursement codes (codes used for insurance or billing), registry outcomes (results from official patient or procedure listings), brain–computer interface (BCI) stability (reliability of BCI devices), continuity audits (reviews of the consistency of care), and reliability drills (tests for system dependability). All these matters are important and need to be addressed with urgency, but first, I must also solve every philosophical question before proceeding. We have a promising future ahead, and with that comes great responsibility about how humankind will walk into it.

Why focus on clear rules when the future looks so exciting? Power without structure causes more mistakes and unfairness, and claims that seem true can be risky if they influence how people live and the laws they follow. Societies that want to last must renew themselves by rotating leaders, shedding outdated policies, tracking access and budgets, and sharing failures so that others can learn from them. Even the best technologies need basics like electricity, water, and engaged people, all merged in respectful communities that seek harmony.

Sarah's grandchildren will not look back in anger or aversion to our time. They will inherit our laws and ledgers, our successes and mistakes, our rails and rituals. If we are careful, they will say we kept our promises: We turned prototypes into public goods. We extended our time without broadening the impact of our errors. We made real choices at life's end and honored refusals. We treated identity as plural without losing parity, powered ambitions cleanly, and tried over and over until we overcame failure to embrace learning from experience.

Welcome to the work of building a longer future. This book will show you key signs, warn you about common pitfalls, and help you set up systems that make longer lives humane. The future remains open, but it can be shaped. Your part starts here.

Part I: The Event Horizon—Understanding the Singularity and Its Immediate Aftermath

Chapter 1

The Last Natural Death

The Forever Generation

Maria sat at the kitchen table of her house in San Antonio with a cup of coffee while the spring sunlight turned the floorboards into neat rectangles. Engrossed in their phones, her grandchildren scrolled through a world of emerging realities. The algorithm randomly showed a video of a sticky song, a notification about a new clinical trial for inflammation, or an article on "good death" contracts.

On the counter, next to the coffee maker, lay a folded program from her brother's funeral. There had been casseroles, calla lilies, and a funeral director who moved with practiced calm. It all felt ordinary. Still, Maria couldn't shake the feeling that something routine had become historic. She wondered if this might have been the last natural death in the family.

She didn't mean that people would stop dying. Maria knew accidents still happened, infections still spread, and wars continued, but she was thinking of the old way of dying, when bodies wore out. Her brother had refused the new treatments, telling her he wanted to "walk the old road." She admired him for that. At the same time, she saw the world changing, with clinics repairing worn-out tissues, dashboards warning of illness before it began, and counselors helping families discuss choices around end-of-life care.

While Maria's generation had learned that death was a certainty, her grand-children had a more complex truth ahead. In the near future, biology would no longer dictate the end of life as often, and this newfound choice would need new rituals, new laws, and new kinds of care.

How Will We Recognize the Threshold?

There won't be a siren to mark the "last natural death." No single paper, device, or law will suddenly make mortality a simple on–off switch. Instead, we'll notice a threshold as clinical, technological, legal, and cultural changes come together in a new, measurable way.

Life Extension in the Clinic: From Promise to Practice

Regenerative medicine is a branch of medical science focused on repairing, replacing, or regenerating damaged or diseased cells, tissues, or organs to restore normal function. It aims to harness the body's natural ability to heal itself by using techniques such as stem cell therapy, tissue engineering, and molecular biology. This field has the potential to treat conditions that currently have limited treatment options, including injuries, degenerative diseases, and organ failure.

This practice often involves growing new tissues in the lab, stimulating the body's repair mechanisms, or using biomaterials to support tissue regeneration. Today, this is not an extraordinary practice but has become routine care.

This has brought along significant changes, as surgeons don't have to choose between heroic measures and hospice when a heart fails or a liver scars beyond repair. Doctors can use autologous induced pluripotent stem cells (iPSCs) to grow patient-matched tissues, and with 3D-bioprinted scaffolds, they can guide the regrowth of vessels, cartilage, and organ subunits. Decellularized matrices enable the reseeding of complex structures that adopt a patient's immunological identity.

Initially, procedures such as these are reserved for patients who are healthy and have single-organ failure. Over the decades, registries reveal surgical success and durable functioning. The practice of regenerative medicine allows for ejection fractions to be restored, bilirubin levels controlled, and dialysis deferred for years. A broader effect is that waiting lists shrink and donor shortages matter less.

Growing Tissues

Let's imagine this scene: In a surgical suite, a clinician carefully holds a graft cultured from a patient's iPSCs. The graft is a small piece of tissue poised to rejuvenate failing organs. Clinicians learn to measure and modulate hallmarks such as cellular senescence, mitochondrial dysfunction, stem cell exhaustion, epigenetic drift, and chronic inflammation for treatments in oncology, cardiology, and geriatric medicine.

Cellular Reprogramming

Senolytic treatments reduce the number of old cells, which release harmful substances and damage nearby tissues. Meanwhile, partial cellular reprogramming protocols are designed to rejuvenate without erasing identity or triggering malignancy.

Clinicians pilot these techniques for specific indications and monitor them for years to ensure safety. They also include support for telomere maintenance, autophagy, and proteostasis as part of standard care. These practices are monitored via special protocols to measure, adjust, and contrast with outcomes.

Somatic Gene Therapy

The first wave of approved gene therapies targeted disorders with single devastating mutations. The second wave goes beyond this and addresses common risk variants that raise low-density lipoprotein (LDL) levels, alter coagulation parameters, or increase amyloid deposition.

The therapy encompasses human germline genome editing (hGGE), a somatic process that is tightly regulated and audited over the long term. However, germline editing remains a red line in most jurisdictions. Regulators and researchers have established biobanks and registries that track off-target effects and late toxicity over decades. As this use becomes more common, we will learn a new vocabulary: delivery vectors, on-target efficacy, immune escape, surveillance windows.

Nanomedicine

Nanomedicine is the technology underlying everything. It entails engineered nanoparticles that deliver drugs exactly where they are needed, with nanoscale sensors placed under the skin to spot shifts in cytokines, metabolites, clotting factors, and tumor DNA at a micro-scale. Other devices dissolve after their work is done.

This technology allows continuous monitoring, which makes early intervention possible: a flare tamped before it swells, a plaque stilled before it ruptures. While this is optional, it will become standard for those at risk and end as an integral part of care.

What Marks a Threshold Rather Than Incremental Progress?

Several highlights will signal the threshold:

- **Improvement in several hallmarks at the same time:** The threshold won't be limited to one drug or device; instead, it will be a combination of regeneration, rejuvenation, monitoring, and targeted editing that will lead to lasting results. Real success will mean fewer strokes, fractures, and cancers in people who receive these treatments, not just better numbers on a chart.

- **Transition from specialized services to widespread access:** Effective treatments will be those covered by insurers and integrated into standard healthcare practices beyond exclusive clinics. Guidelines will formalize them, and residency programs will teach them. There will be a price on malpractice insurers' lists, as well as safety boards that oversee them.

- **Tying biomarkers to hard outcomes:** It is easy to shave years off an "epigenetic clock" and declare victory. It is harder, but necessary, to demonstrate that the clock tracks with real reductions in disease, disability, and death over the years.

This combination of treatments won't make anyone invincible. Instead, it will mean deaths like that of Maria's brother, where organs fail simply because of age, will be much less common. Most deaths will be from accidents, sudden misfortune, personal choice, or rare illnesses that medicine still can't prevent.

Verification Through Population-Level Evidence

Health systems will start to measure aging similarly to other risks, building cohorts, dashboards, and registries that track individuals across decades. Biobanks will link samples with consented data. Researchers will use longitudinal studies to compare longevity care standards with matched controls. This will allow them to adjust the results according to socioeconomic status and access. Governments will promote the use of anonymized data to expedite the advancement of knowledge, while privacy regulators will enforce strict compliance measures. All this will lead to the emergence of a new specialty: longevity epidemiology, trained to separate signal from the excitement of the latest thing.

In parallel, actuarial tables will shift. Health, life, and disability insurers will recalculate liabilities. The world will change, and more 90-year-olds will be able to climb stairs while fewer 60-year-olds will die of heart attacks. As the population's needs change, the government will need to update and redesign pensions and annuities. The map of health conditions, hospice volumes, and causes of death will also shift. In cities, the number of falls from ladders will rise because people still climb them, but traffic fatalities will decline as autonomous systems handle more of the driving.

Digital Continuation: Prerequisites and Verification Regimes

Digital continuation means that a person's mind can be preserved and re-instantiated on a digital device. This idea has been debated for a long time. Whether you believe it desirable or not, it is still possible to measure progress without worrying about the metaphysical implications. Concrete indicators will be needed to build the groundwork and credibility of any claim.

Brain–Computer Interfaces

Brain–computer interfaces (BCIs) are systems that enable direct communication between the brain and external devices. They work by detecting, interpreting, and translating brain signals into commands that can control computers, prosthetics, or other electronic devices. BCIs can be used for various applications, including medical purposes, such as helping individuals with paralysis control robotic limbs that generate stimulation patterns to ease tremors or alleviate phantom pain.

Essentially, BCIs create a pathway for the brain to interact with technology without needing traditional output pathways like muscles or speech. For some time, researchers have been developing experiments with BCIs, and now, those experiments are turning into practices in clinics. The technical curve is bending and showing stable signal quality over the years, better spatial resolution, lower infection risk, and more informative feedback.

Nonetheless, coverage is still partial. Whole-brain access in living humans is not available yet. However, the practical, ethical, and clinical work proceeds with the creation of consent processes, fail-safes, and insurance approval.

Connectomics Maps and Dynamic Models

These other innovations will also scale up. Laboratories will publish complete synapse-level maps of small animals, partial maps of larger ones, and increasingly faithful simulations of defined circuits. These will recreate the following:

- hippocampal patterns that recall

- cortical microcolumns that detect edges

- basal ganglia loops that select actions

A core problem will remain: It won't be enough to capture structure or the map; it will also be necessary to capture dynamics, such as timing, neuromodulation, plasticity, and ongoing learning. Then, researchers will be able to opt for hybrid models that use rich recordings from living brains to constrain computational architectures.

To accompany the process, computing and storage will continue to grow, with caution. Specialized hardware will accelerate neural simulations and model training, but issues related to energy budgets, latency, and error rates will constrain the design. Large projects will create systems focused equally on dependable performance and clear verification.

The teams will propose and test verification frameworks. The central challenge is determining whether a digital continuation is merely a skilled surface-level impersonation. Researchers, ethicists, and families will be interested in proving whether a digital continuation can pass routine quizzes based on the person's known preferences and basic facts about their life. If the person was confronted with a dilemma they once resolved according to a deep conviction, the digital continuation would choose an entirely contrary path.

This exposes the need for verification tests that delve deeper into identity and do not just rely on shallow imitation. Some aspects to consider would be as follows:

- **Consistent behavior under novelty:** Does the digital agent behave like a person in new situations, such as making choices, facing delays, and navigating trade-offs, rather than simply replaying the past?

- **Autobiographical recall with context:** Does the agent quickly gather events, feelings, and meanings, adjusting its understanding when new facts show previous details were incorrect?

- **Value and trait stability over time:** Does the agent exhibit consistent moral values and personality traits, retaining the capacity to learn and change?

- **Relational validation:** Do people who have known the person for decades feel they are interacting with the same person, not just witnessing mimicry? That is, can the agent share private jokes in new contexts, use idiosyncratic phrasings, and make non-obvious references appropriately?

- **Transparent origins:** Is the data chain from acquisition to model documented? Are people obliged to convey express consent that is recorded and revocable where feasible? Can third-party auditors inspect pipelines without violating privacy?

Even with these steps, people will still disagree about whether accurate personal continuity has been reached. The goal of a verification system isn't to solve philosophical debates but to help prevent harm, fraud, and self-deception.

Society Retools for Elective Mortality

Laws, rituals, institutions, and public opinion adaptations will be charted and tracked through legal changes, for instance. In this sense, countries might expand access to medical aid in dying under strict conditions; courts will rule on disputes between family members and digital continuations; and estate law will seek assets when biological life ends but digital presence continues.

Changes in laws and policy follow different rhythms and paths across jurisdictions. For instance, in 2020, New Zealand voters approved the End of Life Choice Act by referendum, with the Act coming into force in 2021. Similarly, the Netherlands, Belgium, Canada, and parts of Australia have several forms of assisted dying with safeguards, while U.S. states differ significantly. These regimes remain contested, and court challenges continue, but what matters for our threshold is not uniformity. Instead, we expect momentum toward treating end-of-life choice as a matter of regulated care, not a societal taboo.

New Rituals and Institutional Responses

Throughout history, each society has developed rituals to deal with death. In a similar way, contemporary and future societies will create new rituals in response to elective mortality. While traditional funerals will remain, "living wakes" will become increasingly common. Beyond disease, families will hold goodbye dinners before death by decision, and communities will develop scripts and songs that talk about grief and autonomy.

Concurrently, clergy will write new prayers and secular groups new vows. In parallel, families will create rites of continuation, welcoming a digital presence to the dinner table, deciding how to introduce that presence to young children, and setting boundaries.

Institutions will actively adapt to a future of longer lifespans and elective mortality. Health systems will integrate care teams focused on prevention, monitoring, and repair for extended lives. Simultaneously, they will maintain essential end-of-life care teams, including those specializing in palliative care, psychiatric evaluation, ethics consultation, and legal processes.

Insurers, on the other hand, will update underwriting practices. Employers will separate career advancement from age, making regular recertification a routine expectation. Additionally, licensing authorities will implement ongoing practice requirements designed to support professional journeys that may extend over decades, including periods of transition and career shifts.

Public opinion, which will become increasingly visible on these issues, will reveal existing disparities that demand action. Across different countries, support for medical aid in dying will show varied trajectories, increasing in some while stagnating or falling in others. This support often correlates with cultural and religious factors, age cohorts, and trust in institutions.

Crucially, the threshold we are examining won't be defined solely by the median view. It will target individuals like Aminatou (discussed below), who will access the care that transforms end-of-life choice from a rhetorical concept into a lived reality. Without deliberate attention to equity, elective mortality might become a privilege available to only a few instead of a universal right.

From Threshold to Meaning

The threshold matters because it changes the kinds of questions we must answer. Some of the clinical questions will become:

- Which combination of interventions will do more good than harm for this person with this genome and this history?

- When is restraint the best medicine?

The legal questions will become:

- How do we protect against coercion while respecting autonomy?

- What counts as informed consent when choices are new?

And the cultural questions will become:

- How do we honor the option to continue and the choice to end?

- How do we keep attention on those who cannot access either safely?

A Brief Historical Lens

No human culture has lived without rituals for death. The Ancient Egyptians mummified and mapped the afterlife with confidence; the Greek Stoics taught the importance of preparation; medieval Europeans depicted dances of death and developed elaborate mourning customs; and many Indigenous peoples wove their ancestors into daily life through offerings and stories. Modern medicine has extended lifespans and moved death into hospitals, sometimes out of sight of children and neighbors, but 20th-century wars and pandemics have brought it back into public view.

When the moment of the threshold comes, it will be different, not because death matters less but because its meaning has changed. When biology can be mended and extended, communities will seek answers:

- When do we stop?

- Who decides?

- How do we mourn a choice?

- How do we coexist with a loved one who is present digitally but physically absent?

The old rituals will not be wasted; they will provide foundations for new ones. The old ethics will not be obsolete; they will be tested against new fears.

Four Lives at the Edge

The following stories depict different scenarios of emerging dilemmas and ongoing challenges. Each of them sheds light on unique aspects of the threshold, and they have specific policy and legal consequences.

Oviedo, Spain, 2041: Miriam Esquivel, 88

Miriam lives in the house she and her husband built in the 1970s, with grapevines trained along a low wall and the soil dark after rain. When her cardiologist offers a series of rejuvenation treatments that would likely restore her stamina, she declines.

She prays, she cooks, she sits in the sun. Her children are devastated yet respectful; some friends are proud; strangers on the internet argue. As she reflects, Miriam wonders silently, *Will my grandchildren still plant vines, or algorithms?* Her death is ordinary, beautiful, and challenging.

Policy consequence: Legislators in her region pass explicit language protecting the right to refuse longevity care without losing access to palliative services, and they fund the teams that support that choice. Autonomy becomes practical, not performative.

Dakar, Senegal, 2039: Aminatou Mbaye, 34

Aminatou works in a crowded tech hub and returns each night to a neighborhood where the lights go out when the grid hiccups. A rare genetic condition is accelerating her biological age. The hospital can offer pain control but not the cellular therapies she reads about. The lottery for treatment spots is a thin cover for scarcity. Her friends fundraise and petition; the Ministry of Health is slow, then faster. She dies. The city lights candles, and not only for her.

Policy consequence: A national fund is created to subsidize life extension for people under a certain income threshold. The government adds data collection requirements to ensure eligibility is enforced without stigma. They also search for international partners to increase capacity so that lotteries become scheduling tools instead of rationing devices.

Singapore, 2040: Theo Sun, 46

Theo collapses at work as a thin ripple of electrical confusion crosses his heart. The building's medical AI flags an emergency, and his wearables call a public ambulance. The nearest hospital is notified and preps a team.

Somewhere between signal A and signal B, a translation module introduces a delay. Theo dies before the second defibrillator shock. The city is stunned. Auditors trace the failure through software and regulation, producing a report that is so clear that even nontechnical citizens can follow it.

Policy consequence: Life-critical systems are required to carry redundant channels, perform independent audits, and carry out drill schedules. Safety and liability are attached to networks, not just devices. It becomes harder to certify "smart" without proving "safe."

Paris, 2042: Professor Emilia Favreau, 79

A cultural historian who spent her career writing about mourning in medieval Europe, Emilia has been forthright about her decision to refuse life extension. Her final lectures are packed, she writes a short book titled *Choosing to End*, and she donates her papers to an archive that will be shared with future students, and, if her family agrees, her digital continuation. The night before she dies, her extended family listens as she tells stories they have heard, as well as some they have not.

Cultural consequence: Universities add mortality studies to general education. Hospitals and hospices partner with artists and clergy to design rituals that neither resemble imperfect copies of older customs nor pretend everything is novel. Families have models to follow when their own time comes.

The Politics of Consent and Coercion

As end-of-life options become normalized, new risks will appear. Families may pressure elders to choose an exit that seems "efficient." Insurers may be tempted to steer patients to less expensive options. Conversely, tech companies may market continuation and rejuvenation as moral duties to increase profit: "Don't abandon your family; don't abandon your work."

Protection requires law, but it also requires culture. Societies will need to honor both options: to stay and to go. Nobody should be pushed or persuaded toward either.

The process of decision-making will need a close look to ensure health systems develop layered safeguards that include second opinions, mandatory waiting periods, psychiatric evaluations, and trained counselors who can distinguish between ambivalence and coercion. This way, consent will be treated as a process. For digital continuation, consent will become even more complex:

- Will a person's agreement extend to future updates of the model?

- Will consent be withdrawn?

- Who will decide what happens when a person's digital presence diverges from promises made while biological?

Measuring What Matters

Communities often create their own stories. The only way to know if elective mortality is becoming more real and humane will be to measure it. These new dashboards may be visually unappealing. They will reveal which neighborhoods have access to longevity care and which do not, which families utilize end-of-life options with care and which with pressure, and where digital continuations seamlessly integrate into family life or cause tension. When the public is willing to face these facts and adjust budgets and laws, that's an absolute threshold.

What the Threshold Is Not

This shift about the end of life isn't about immortality. People will still die from accidents, from illness, or by choice. The real point is having agency, not living forever. Longer lives aren't always better lives. For each personal story, extra years can be a gift or a burden, depending on health, purpose, and community. The most dangerous ideas are the simplest ones: Death means failure, continuing is an obligation, or choice is always easy.

What Comes Next

We still have to discuss three pillars (AI, biotechnology, and neurotechnology), which must reinforce one another to make elective mortality broadly available and safe. The consequences will open new questions: who we become when timelines stretch without the forced resets of generational turnover; how economies and polities adapt when experience compounds; what love, memory, and meaning look like when time keeps giving us tomorrows.

For now, it is enough to imagine Maria's grandchildren coming home from another ordinary funeral and knowing that they have witnessed something their children might rarely see.

Closing: New Options

If the last natural death happens quietly, it will be because people have spent years talking about how to live with choice. They will have learned new ways to speak and overcome old shame. They will have found a way to thank biology for the time it gave them and to welcome a longer, stranger future shaped by invention and care.

Maria will have taught her grandchildren to cook, and they will have taught her how to use a smartphone. Together, they will do what humans have always done: look at an unexpected future and find their place in it. But this time, they'll have the option to choose when to leave.

Chapter 2

Convergence Point

In the previous chapter, we wondered how we would recognize a threshold. The answers pointed at elective mortality becoming a genuine concern when clinical outcomes can be recorded and measured. Other aspects to keep in mind are verification systems, population-level evidence, and legal changes that signal death as optional.

This chapter explains why that threshold is realistic and what it will take to cross it. The argument is simple: No single technology will make the shift happen. Diverse scientific fields will need to cooperate: AI to accelerate discovery and design; biotechnology to repair and modify bodies while connecting them to machines; and digital continuation to capture, preserve, and restore a person's functional patterns. Each pillar will bring something the others need. Together, they will create a system where feedback shortens cycles and reduces errors. The more these fields progress, the better able institutions will be to select and adopt what works.

Why Convergence Is the Only Useful Lens

Convergence describes the causal structure beneath virtually every headline you have read in the last decade. The advances that will enable the threshold rely on the indispensable cooperation of different fields.

Biotechnology generates vast amounts of data, like single-cell omics, long-term imaging, and continuous physiological streams. This data is not helpful without AI to find patterns and suggest ideas. Those ideas then need to be tested in laboratories, delivered in forms that can be manufactured, and monitored for safety over a long period. Without this process, AI discoveries will stay as papers and press releases. With it, they can become real treatments that change what insurers cover and what doctors do.

For instance, let's think about the celebrated acceleration of protein structure prediction and interaction modeling. This method combines advanced computational techniques and algorithms to speed up the process of determining the 3D structure of proteins and how they interact with other molecules. It replaces traditional methods like X-ray crystallography and nuclear magnetic resonance (NMR) spectroscopy, which can be time-consuming and expensive. Instead, computational approaches, including machine learning, deep learning, and molecular simulations, enable faster prediction by analyzing protein sequences and existing structural data. This acceleration helps in understanding protein functions, supporting drug design, and studying biological processes more efficiently.

AI systems have compressed a bottleneck that had frustrated biologists for years. These scientists have traditionally focused wet-lab resources on higher-probability candidates. Recently, lab validation times have decreased by 50%, enabling scientists to rapidly iterate designs and focus on potential breakthroughs.[1] However, what makes these tools most useful is that labs can validate them, manufacturers can produce them at scale, and regulators can establish trial pathways. The real achievement was not an algorithm but a braid of machine learning, bench science, and governance.

Another good example is the intent to decode speech and motor functions through BCI studies. Gains in decoding quality depended on the use of better implant materials, improved surgical techniques, and signal processing, which produced cleaner and more stable data. Algorithms trained on that dataset developed interpreters that effectively restored specific patients' lost functions. However, the real-world impact occurred when institutions adapted rehabilitation protocols and provided paid compensation for clinical deployment. Again, the focus of the story is not a single field; it is the interaction of AI, biotech/engineering, medicine, and policy.

These examples show a pattern: On its own, each pillar can create impressive demonstrations; together, they lead to real-world adoption.

The Three Pillars and What They Do

Below, we'll consider the pillars one by one to better understand what they bring on their own and how they relate to the others.

Artificial Intelligence

AI in this stack is not an abstract "thinking machine." It is a set of tools that translates messy, high-dimensional biological and behavioral data into testable hypotheses and candidate interventions. These tools search and optimize across vast design spaces, from small molecules to delivery vectors to electrode geometries and stimulation waveforms. In addition, they interpret and compress neural and physiological signals into models that clinicians and researchers can use.

AI's strengths are pattern discovery, simulation, and search. Despite this, the risks are familiar to anyone who has deployed models in production: causality and domain shift, bias and opacity, energy and carbon costs, and the temptation to mistake statistical compression for understanding.

Consider the scenario of a patient whose condition is mis-triaged due to an opaque AI model, an oversight that delays treatment until it is nearly too late. This is a realistic scenario that reveals the moral urgency of ensuring our systems are transparent and auditable.

To address AI risks comprehensively, it is crucial to implement concrete risk mitigation strategies. These strategies should include the use of model interpretability tools that help demystify black box algorithms so clinicians can understand the decision-making process of AI systems. Other strategies include regular bias audits to identify and mitigate potential biases in datasets and algorithms, and energy-efficient AI models that can significantly reduce their carbon footprint. Lastly, creating feedback loops and audit trails ensures continuous improvement and accountability, particularly in critical areas such as healthcare.

The meaningful standard for AI here is not a stylized benchmark. It is pre-registered, independently audited clinical trials showing that AI-guided protocols improve hard endpoints, such as event-free survival, hospitalization-free days, and functional recovery, relative to standard care. It needs to be robust under domain shift, and there must be a transparent, versioned pipeline from data acquisition through deployment that a third party can reproduce. If you are an engineer or product manager, this will sound like standard practice. That is the point. The same discipline that keeps ad systems or logistics platforms from degrading or discriminating applies even more strongly when you are triaging a stroke or adjusting a spinal stimulator.

Biotechnology

Biotechnology's role is to modify and repair biological systems, as well as to build the interface layer between biological systems and machines. This includes a wide spectrum:

- somatic editing and control of gene expression

- cell, tissue, and organ engineering

- delivery vectors and biocompatible materials

- continuous sensing platforms that link detection to pre-emptive intervention

In aging biology, insights about senescence, proteostasis, autophagy, and mitochondrial function flatten disease and disability curves when practitioners apply them in interventions. These interventions must meet stringent safety and efficacy standards before being widely adopted.

For example, CAR-T cell therapy is a treatment that has shown remarkable efficacy in targeting and eliminating cancer cells by modifying a patient's own immune cells. This therapy embodies the convergence of biotechnology with clinical practice. Its benefits can already be seen in life-changing treatments for specific conditions and in the fast progress of new technologies like organoids (miniature organs grown in labs), scaffolds (structures that support tissue growth), and 3D bioprinting (printing living tissue).

Besides ensuring safety metrics, institutions must ensure reimbursements and access so people can effectively reach these benefits and equity is preserved. These types of treatments will require large-scale manufacturing and facilities to supply a growing population. Thus, governance within each country will need to foresee budgets and logistics to ensure equitable access.

The standard for progress in biotech is not a spectacular before-and-after biomarker. It is a reduction in clinically significant events and mortality throughout many years. The records to consider are gathered in cohorts under "stacked" longevity care. The results are relative to matched controls, tracked in public registries with adverse event and equity metrics.

Health systems code the information for patients and healthcare providers to gather and organize information. These records showcase the procedures that prolong organ function or the somatic edits that can reduce lifetime risks due to a cardiovascular event. The following decade will bring the fruits of this surveillance, showing that safety improves, disparities shrink, and protocols are refined.

Digital Continuation

We are using this term to emphasize practices of continuity and verification that extend a person's existence or consciousness by transferring, preserving, or emulating their mind, memories, personality, or identity in a digital or machine-based form. This "digital continuation" could allow a person's experiences, thoughts, and awareness to persist beyond their biological lifespan.

Let's go beyond metaphysical claims about the nature of consciousness. We don't say that machines are conscious. It is more concrete than that. The starting point is that a digital agent is presented as the same person who consented, was treated, promised, or parented. With this in mind, we wonder: *What must be true about acquisition, modeling, and verification so that institutions, families, and the law can act as if that were the case?*

This goal still remains distant, but there have been solid advances toward achieving it. BCIs in humans can restore specific functions, such as communication and motor control, while also enhancing stability, and hybrid models fuse recorded dynamics with learned architectures to simulate specific circuit-level behaviors.

On the governance side, the main efforts should target delivering proposals for provenance chains, third-party audits, and bounded-purpose use. Experts are taking theory into pilot practice.

In this context, there is a critical point: Some marginalized voices demand continuity verification. We must recognize whose histories are most at risk of erasure if provenance chains fail, and we must aim to emphasize the importance of inclusive practices devoted to preserving those voices and histories.

Besides these institutional policies, the whole community must commit to cooperating in creating audit boards through socially responsive practices. Governments and communities can take several pragmatic steps: engaging marginalized communities from the outset of a project, ensuring transparency through regular updates and feedback sessions, and establishing clear guidelines for participation.

A practical checklist for inclusive governance might include the following:

- outreach strategies to connect with community leaders

- accessible communication channels for ongoing dialogue

- provisions for fair representation on audit boards

- mechanisms for community-led decision-making

These practices, which ensure community engagement throughout the process from the beginning, might help institutions work toward equitable governance structures that genuinely reflect and protect diverse community interests.

The limits to the scope of application of digital continuation are equally clear. As with many social studies, static "maps" are insufficient. Policymakers and specialists must recognize when to capture and infer timing, neuromodulators, plasticity, and ongoing learning, since factors such as energy, latency, and error budgets will become increasingly significant as systems scale.

Besides these social implications, there is another aspect to consider at the individual level: Digital continuation opens many ethical debates about the risks to a person's existence beyond biology. Consent and provenance must work at scale and be revocable where feasible. Moreover, legal frameworks and cultural practices must deter and punish impersonation and exploitation. The social and legal status of digital agents will be contested and will depend on contextual conditions. In this aspect, progress will not be a single test passed but an ongoing process.

Institutions will continuously adopt standardized verification regimes that go beyond imitation. The tests will cover the following aspects:

- behavioral concordance under novel conditions

- autobiographical recall with context and appropriate emotional coloration

- value and trait stability across time windows and learning conditions

- relational validation by long-time intimates in ways that resist mimicry

- transparent provenance from acquisition through model updates, so a third party can audit without violating privacy

These criteria, which were introduced in Chapter 1 as grounding principles, now turn into effective practice.

How the Pillars Reinforce One Another in the Real World

Once you see the stack, you also see the loops that bind it. The discovery and design loop is the most familiar:

1. Single-cell and longitudinal data produce hypotheses.

2. AI systems nominate targets.

3. Clinicians and biologists use editing or cellular interventions to probe those targets.

4. Wet-lab readouts and safety analyses feed back into the model.

5. The next iteration refines the intervention.

To illustrate this, let's consider a patient named Jane who participated in a clinical trial for a novel therapeutic intervention. Initially, the treatment she underwent showed moderate success. The medical team used refined models and AI-driven insights, and after the third iteration, Jane experienced significant improvement in her symptoms and quality of life. This process shows that preregistration to keep track of adverse events and external replication helps in refining interventions.

The interface loop is equally important. Better materials, sensors, and surgical techniques produce cleaner signals for BCIs and other implantable devices. The result is cleaner signals that improve AI decoders, which lead to many positive outcomes: lower-power and more precise stimulation that broadens indications and reduces side effects.

Payment encourages real-world data collection, which creates ongoing information that helps to improve future models and guides regulators to make better rules. There are clear limits to this process. For instance:

- ensuring the device works with the body without causing harm or infection over many years

- managing the device's use and security

- making sure the benefits reach the whole community and not just the rich areas or top research centers

The modeling loop for digital continuation is the least mature but the most sensitive. Detailed in vivo data refines essential models, while connectomics establishes a framework; AI compilers and dedicated hardware then translate these models on to platforms optimized for energy efficiency and dependability. Yet, the ethical and societal challenges intensify. Assertions about continuance carry significant implications. That is why verification regimes must be standardized, preregistered, and audited. The goal is not to engage in a philosophical debate but to reduce harm, fraud, and self-deception while enabling families, clinicians, and courts to reason coherently about continuity claims.

As we have already pointed out, there is also a population loop that often goes unremarked in technology writing. As stacked longevity care reduces acute events and extends healthy years, actuaries and insurers need to revise and update their models. Institutional changes reallocate budgets and attention. The result is positive feedback, because what was once rare becomes routine, and experiments become part of the curriculum requirements. For builders, for instance, this loop should be near the top of the risk register. A useful road map should explicitly include reimbursement codes, credentialing, and training.

To guide this process, consider these key road map checkpoints:

1. Secure reimbursement code applications to ensure insurance coverage.

2. Develop credentialing programs for healthcare professionals.

3. Implement clinician training to integrate new technologies and protocols into practice.

4. Establish partnerships with regulatory agencies to smooth the approval process.

5. Create feedback mechanisms for the continuous improvement of protocols and devices.

Certainly, the details of planning and budget will depend on the immediate context, but these guidelines serve as a general starting point.

Recognizing Convergence in the Wild

Convergence is a system of loops; it announces itself in patterns that go beyond metaphysical disputes. In clinics, it is possible to see that regenerative medicine has become a protocol rather than a privilege. The integration of senescence modulation, proteostasis support, and autophagy-promoting regimens into standard care suggests that interventions are shifting.

Neural interfaces send signals that last for many years and work well in medical settings. A BCI that helps a person who cannot move or speak to communicate or control devices, and works safely for years, is more valuable than one that reads data faster but only works for a few days, unless the faster system can also handle the body's immune response, power failures, and everyday challenges.

It will be important to track and check the history of the devices, data, and models used. There should be ways for patients to get the costs covered by insurance or other programs, and these should expand over time. Without these things, we're just talking about interesting science. With them, it becomes real care for real people.

For digital continuation, we will acknowledge the landmarks when institutions adopt verification regimes and provenance practices as part of their routine. We will continue to disagree about subjective continuity. We should expect the first applications to be narrow and bounded, focusing on continuity for specific legal choices, relational commitments, and constrained forms of work. The wider the claim, the tighter the verification should be.

A hype filter for convergence is simple to use and hard to fake:

- Do not count style-imitation benchmarks dressed up as "understanding."

- Do not count single-site, underpowered clinical studies that are unregistered and unreplicated.

- Do not count "Turing-like" chats as evidence that a system will out-perform therapists or that a digital agent is the person your family lost.

It should rely on the practices listed below:

- Count preregistered trials, independent audits, longitudinal registries, reimbursement codes, curriculum changes, and actuarial shifts at major carriers.

- Count device reliability over the years.

- Count failure-mode audits and rollback plans.

- Count transparent dashboards that track access and safety by geography, race, and income.

- Count the corrective funding when disparities widen.

To select the most relevant metrics for a project and ensure the expected outcomes are achieved, these steps should be followed:

1. Start by identifying the project's primary goals and the stakeholders involved.

2. Select metrics directly related to those goals; they should measure impacts that stakeholders deem critical.

3. Consider the stage of development your project is in. Early-stage projects may prioritize metrics such as innovation potential and research milestones, while mature projects may focus on scalability and adoption.

4. Tailor metrics to address specific industry standards and regulatory requirements to ensure compliance and relevance.

As we move forward, I encourage you to consider which specific metric you will focus on measuring this quarter. Moreover, you might reflect on how you will participate in ensuring accountability and meaningful progress in your field. By taking collective responsibility, we can transform cautious optimism into community-driven change.

If you follow AI, you will recognize many of these themes. Generalization across domains is meaningful only when preregistered protocols are independently replicated and tasks have human-relevant endpoints. An agent that autonomously generates hypotheses that separate lab tests and that improve outcomes in medicine or materials would be a genuine step, whereas "emotional Turing tests" tell us mostly about our willingness to project mind into pattern.

In biotech, we need to replace headlines about "biological age reversal" with measured improvements in disease, disability, and mortality over time. Reports that a clone "fooled its family" are less interesting than a study in which families consented, auditors supervised, and reconnection unfolded under rules that prevented abuse.

Quantum computing deserves a separate paragraph of optimism, yet it must be approached with prudence. Specific subproblems, such as certain classes of simulation and optimization, may eventually provide benefit. However, today, the error rates, decoherence, limited qubit counts, and maturity of classic baselines suggest caution. We must treat quantum computing as a targeted research track, not a force multiplier to include in our road map. Teams that plan to rely on quantum acceleration across the board are preparing for a world that does not exist yet.

Ethics and Equity Are Embedded at the Foundation

Convergence increases leverage, and if it is left to drift, it also amplifies inequity. Community participation should govern biobanks and training corpora to avoid "precision inequity." Models work best on populations that supply the most data and fail where they are needed most.

Also, communities should be involved in the creation of protocols and data use agreements. Documents that establish legal frameworks should include the right to withdraw consent where feasible. Consent must be treated as a process rather than a one-time signature, especially for the collection and use of neural data and the continuation of claims. Consent should be revocable where feasible, documented with provenance, and bounded by purpose. Family members should not be able to override explicit directives; companies should not be able to lock in rights to a person's data "in perpetuity."

On the other hand, societies can take preventive measures to avoid early monopolies of the benefits. For instance, communities can work on programs that outline coverage and pricing models for longevity care. Transparency on access by geography, race/ethnicity, and income should be a requirement for receiving public funds. When gaps widen, budgets should move.

Finally, we should treat energy and reliability as ethical constraints. Compute budgets, power use, and fault tolerance are not isolated engineering issues. They determine who can benefit and at what cost. A device that fails during a storm is not reliable; a continuity agent that fails when a data center experiences an outage is not trustworthy. Engineering plans should consider degradation and redundant paths and account for carbon as part of their success criteria. If you are in policy or procurement, don't hesitate to write those expectations into contracts and licensing.

What to Expect Next

Convergence, as we have explained in this chapter, is the architecture for ever-lasting life. In the rest of this book, I will share the plan and the integration tests. First, I will explore the biological dimension. There are organisms that already bend time. In nature, hydras, planaria, tardigrades, long-lived trees, and corals have developed mechanisms that underpin their resilience. These mechanisms include vigilant DNA repair, open-ended regeneration, dormancy, and stress responses, and they are the biological scaffolding beneath the stacked longevity care we're exploring.

Then, I will shift the focus to identity. It is possible to assume that, once neural interfaces are stable and identity governance has become practical, pressures to copy, branch, or continue will arrive sooner than most expect. The spotlight will be set on household-scale design patterns for branch registries, continuity trusts, and presence-weighted obligations. At this point, we don't need to devote ourselves to metaphysical debates. Instead, those patterns must uphold their ethical responsibilities toward individuals who decline enhancement or continuation, demanding equal legal rights for both individual and collective identities.

Next, we will learn how to search for the signals you can watch for in cities, economies, and governance for populations without end. If you work in urban planning, macroeconomics, or civic tech, these sections speak your language; they cover post-growth metrics, circular resource flows, rotating leadership and sortition, AI-assisted but human-accountable decision systems, and institutional designs that resist ossification. The core argument here is that convergence is the substrate and stewardship is the constraint.

The final stretch turns inward to the psychological realm. It explores the cognitive and emotional realities of decision-making in a world where time no longer forces closure, as well as the rituals and agreements that allow relationships to be renewed rather than assumed. If the previous parts describe rails and trains, this last one describes passengers finding meaning in motion.

Closing: Convergence as a Work Plan

Convergence is not a brand; it is a plan for turning prototypes into public goods. The progress you have seen in discovery, design, interfaces, modeling, and policy only happens when we also invest in the systems that support them: verification, consent, reimbursement, equity, and energy-aware infrastructure.

If you are in your late 20s or 30s and working in AI, biotech, or neurotech, you will help decide whether the next 20 years will be marked by demos or lasting change. Insist on preregistration and audits. Write consent forms as if they were for someone you care about. Measure real outcomes and population changes. Build for reliability and low carbon. Budget for training and access, not just intellectual property and conferences. This is how you help create a future that works for everyone.

So far, we have set the stage for a path into an everlasting life that we are already walking. We have also outlined a road map to analyze what we will face and what we must consider. Let's start unraveling what nature can teach us about bending the limits of time and matter.

Part II: The Biology of Forever—Biological Precedents for Immortal Life

Chapter 3

Earth's Immortals

What Biological Immortality Means

When biologists apply the phrase "biological immortality," they are not claiming that a creature cannot die. To make this concept more relatable, let's consider how humans face growing health risks as we age. Each birthday incrementally tilts the odds toward incidents like heart attacks or strokes.

In contrast, the phrase "biological immortality" identifies organisms with no detectable age-related increase in mortality risk under stable conditions. Death still comes by infection, predation, accident, starvation, or catastrophic stress. What appears to be different is that the failure rate does not slow down sharply as damage mounts and the ability to recover diminishes. "Immortality," then, is a statement about risk and renewal, not a promise of eternal existence.

Many species have characteristics that give them biological immortality. Hydra kept in benign laboratory environments are the iconic case. Their death rates stay the same over several years, showing little to no aging effect.[1] Certain jellyfish can reset their life cycle to a juvenile stage. These sea creatures avoid an obligatory terminal state, although they die readily in the wild from ordinary hazards.[2]

Biological immortality is found in animal and vegetal species, and it also entails adaptability. Trees and corals endure not by keeping a single fragile module perfect forever but by compartmentalizing damage and sustaining living, repairable tissue around a durable scaffold.

These examples don't provide a recipe for human immortality. However, they are important because they reveal diverse strategies that biology has evolved to extend functional time:

- continuous replacement

- damage suppression

- reversible state changes

- modular architectures that tolerate local failure

Why study these unusual cases? They help us see what is possible when DNA is well protected, proteins stay in order, aging is rare or quickly fixed, and body structures support long life. Each approach offers lessons, but not all can be applied directly to humans, and every benefit comes with trade-offs. For example, the longest-lived mammals have tissue structures that resist cancer, but these structures are very different from ours.[3]

This chapter examines key organisms, outlines their primary strategies, and then focuses on human pathways where the science is well-established. I explore some of Earth's most resilient organisms. Each example reveals unique strategies for survival and provides valuable insights into potential human applications. As you explore these remarkable cases, you will uncover lessons that could inform innovations in human longevity and health.

Tardigrades: Survival by Shutting Down

Tardigrades, also known as "water bears," are tiny eight-legged animals that can do something startling when desiccated or otherwise stressed: They enter a "tun" state, a form of cryptobiosis in which their metabolism drops to levels that are hard to detect. In this state, tardigrades endure extremes that would shred ordinary cells. They can tolerate near-vacuum conditions, large temperature swings, radiation exposure, and the absence of liquid water.

Two results anchor what matters here. First, in a clever cross-species experiment, researchers identified a tardigrade-specific protein, Dsup (for "damage suppressor"). This protein binds DNA and reduces X-ray-induced damage when expressed in human cells.[4] This is a molecular foothold: a specific protein that measurably increases DNA protection under ionizing stress.

Second, another experiment tested tardigrades' survival traits in space. Dehydrated tardigrades survived brief exposure to the vacuum and solar ultraviolet radiation of low Earth orbit during a European Space Agency experiment. When rehydrated, a fraction revived and reproduced.[5] They were not "living" in space in an active sense; they were paused, protected by a suite of adaptations, and capable of restarting after insult.

The greatest lessons here are that biology can operate in extremely low-damage modes, and that molecular chaperones and DNA-binding proteins can significantly alter the damage equation. Humans will not become cryptobiotic, but our cells and therapies already take advantage of low-damage states. We can mention, for instance, hypothermia in cardiac arrest, vitrification in cryopreservation, and carefully dosed radioprotectants in oncology. This suggests that rationally engineered radioprotection may be possible in narrow clinical contexts.

Hydra: A Body Plan Built on Constant Renewal

Hydra are simple freshwater cnidarians with a diffuse nerve net and a tubular body capped by tentacles. Their claim to fame is their relentless capacity for tissue renewal. Three stem cell lineages continuously produce the epithelial and interstitial cells that make up the animal.

Under laboratory conditions that eliminate extrinsic mortality, hydra populations exhibit negligible senescence, indicating that their risk of death does not increase significantly over time.[6] Cutting a hydra produces hydra again; feeding and temperature modulate their growth, but there is no obvious intrinsic endpoint forced by failing renewal.

To remain attached to the truth, we need to make two clarifications. First, hydra are very different from vertebrates in terms of both architecture and complexity. Thus, we should not infer that what is easy for a hydra will be easy for a human. Second, "negligible senescence" is not a claim about centuries-long individual lifespans; it is a claim about flat risk across years in a benign environment.

Even so, hydra exemplify a principle that scales: When damage is constantly diluted by replacement and when renewal machinery stays robust, actuarial curves can flatten. Human tissues that naturally turn over quickly, such as blood, gut epithelium, and skin, offer a partial echo of this strategy. Our task is to maintain the vigor and fidelity of their stem and progenitor compartments rather than accepting decline.

Planarians: Regeneration on Demand

If you cut a planarian flatworm into pieces, each piece can grow into a complete worm. That astonishing feat is powered by a population of clonogenic neoblasts: adult pluripotent stem cells that can produce every tissue.[7] They have positional information along the body axes that guides patterning, ensuring that heads develop only where they belong. Researchers are exploring how gradients, transcriptional networks, and bioelectric cues coordinate this outcome.

Planarians in nature are not actuarially immortal since they live under predation and environmental stress. However, they demonstrate that adult organisms can maintain a deep reservoir of regeneration and deploy it repeatedly without exhausting the pool.

In mammals, regeneration is tightly regulated. The mammalian liver can regrow, and skin and blood replenish, but complex structures like limbs do not reappear. Planarians redirect our focus toward specific environments and molecular cues that enhance or hinder the capacity for tissue regeneration. This is helpful in analyzing anticancer strategies in humans, such as tight cell cycle control and pro-senescence signaling, that may lock in repair deficits that a short-lived animal can tolerate but we cannot.

The Immortal Jellyfish: Life Cycle as an Escape Hatch

Turritopsis dohrnii is the species behind the "immortal jellyfish" headline. It can reverse its development: Under specific stressors, the free-swimming medusa regresses to a juvenile polyp through a process of transdifferentiation, in which mature cells switch their fates.[8]

In laboratory settings, this cycle can repeat multiple times. In the wild, jellyfish are still subject to ecological reality, and therefore predators and pathogens end most of their lives. Still, the developmental reset is biologically profound. It tells us that the barrier between differentiated states is not absolute in all animals, and that controlled fate-switching can allow an organism to abandon a failing adult body plan and start over.

In mammals, the closest analogs are induced pluripotency and partial reprogramming; in these technologies, transcription factors (the "Yamanaka factors") reset epigenetic markers that correlate with cellular age. Strong reprogramming can erase a cell's identity. But using short tissue-specific doses in mice has helped restore function and reversed some signs of aging without causing problems. This works especially well in retinal ganglion cells, which are nerve cells in the eye used to restore vision after damage from nerve injury or glaucoma.[9] Immortal jellyfish don't show us how to do this in a human; however, they indicate that, in principle, a mature organism can invoke programs that rewind.

Naked Mole-Rats: Mammals That Age Differently

The naked mole-rat stands out to researchers in the field of aging because it defies expectations. These mammals live in low-oxygen environments, can survive for decades with little sign of aging, remain fertile well into old age, and rarely develop diseases such as cancer.[10]

One reason is that their tissues contain a special form of hyaluronan, which helps prevent cancer.[11] They also keep their proteins in good shape, have unique fat profiles, and respond to stress differently than short-lived rodents.

The main lesson is that long-lived mammals don't rely on a single change. Instead, they adjust multiple systems simultaneously, such as tissue structure, repair, and immune defense. For humans, this means that a combination of small, coordinated steps will likely be more effective than any single big intervention.

Ancient Trees and Corals: Endurance by Design

Trees like bristlecone pines (*Pinus longaeva*) are among the Earth's longest-lived individual organisms, with living specimens over 4,800 years old.[12] They don't accomplish this by keeping a single vulnerable component perfect forever. Their secret is that they concentrate living function in a thin ring of cambium beneath the bark and let the interior wood become inert support. They grow slowly, allocate precious resources to maintenance, compartmentalize damage so that pathogens and decay are contained, and utilize chemical defenses to discourage invaders. Giant sequoias employ a different tactic: Thick, tannin-rich, fire-resistant bark and massive architecture, alongside slow growth, make them hard to kill.

In the depths of the ocean, corals exhibit a modular strategy at various stages of their development. Massive *Porites* colonies can persist for centuries as genetically identical polyps lay down aragonite skeletons, one layer at a time. When local damage occurs, living tissue grows over the wound, and when polyps die, their neighbors encrust them and continue the edifice.

Coral structures act as natural archives, their growth rings capturing detailed records of oceanic changes year by year. Meanwhile, coral reefs continually maintain their form through ongoing small-scale restoration processes, functioning as dynamic, self-sustaining ecosystems.[13] The general lesson is that design matters. Distributed living tissue around an inert scaffold, compartmental boundaries that stop damage from spreading, and modular growth that supports local replacement are engineering choices, not simple biological facts.

Mechanisms That Matter: A Synthesis

Seen together, these organisms outline a small catalog of strategies for resisting the passage of time. To prioritize these strategies for practical human applications, we can categorize them into three main levels, primary, secondary, and tertiary defenses, supported by complementary defense levels.

Primary Defenses

Vigilant DNA integrity represents a primary defense. Tardigrade Dsup is a vivid example of a protein that reduces ionizing radiation damage by physically associating with chromatin. Long-lived mammals, meanwhile, display shifts in DNA repair priorities, checkpoint robustness, and apoptosis thresholds that favor the maintenance of genome stability without compromising growth.

The principle is simple. The aim is to reduce the rate at which irreparable lesions accumulate. As a consequence, the downstream cascades that force senescence or transformation will be rarer.

Secondary Defenses

Protein maintenance and cell cleanup serve as secondary defenses. Many long-lived organisms are adept at maintaining the shape of their proteins and the functionality of their cellular components. They use systems that remove damaged pieces before problems build up.

In mammals, drugs (such as mTOR inhibitors and AMPK activators) and lifestyle changes (such as exercise and specific diets) have helped animals live longer and improve certain tissues. The application of these ideas to humans is still being studied, but the key takeaway is that regular maintenance is preferable to letting problems accumulate.

Tertiary Defenses

The suppression or management of cellular senescence acts as a tertiary defense. Senescence is a double-edged sword: It prevents dangerous cells from
dividing, yet senescent cells secrete inflammatory factors that corrode function
in their surrounding areas. Systems that prevent senescence, clear senescent
cells efficiently, or constrain the inflammatory secretory phenotype avoid the
self-reinforcing inflammatory state often seen with age.[14]

Hydra avoid senescence through constant replacement, while mammals, on
the other hand, must balance clearance and suppression with cancer prevention.
Science has developed senolytic drug candidates. These agents selectively kill
senescent cells in mice and are being tested in early human studies for fibrotic
diseases.[15] This places this lever on the translational map, accompanied by appropriate warnings about off-target effects and long-term risks.

Regeneration capacity also falls into the tertiary defense category. Hydra and
planarians can rebuild complex anatomy, while mammals typically cannot. The
mammalian push to regain some of this capacity has two branches. One uses
exogenous cells, scaffolds, and organoids to transplant replacement parts; the
other attempts to rejuvenate and mobilize endogenous cells by manipulating
niches and epigenetic states.

Complementary Defense Mechanisms

Controlled transient exposure to reprogramming factors in mice has reversed
specific functional deficits. However, identity erasure and tumorigenesis remain real risks when the same tools are misused.[16] The key here is precision:
tissue-specific, transient interventions governed by robust safety monitoring.

Metabolic tuning and stress responses straddle the distinction between secondary and tertiary defenses. Both tough invertebrates and long-living mammals have ways to handle stress that help them avoid the problems caused by
going dormant or aging too fast. Their bodies use proteins that respond to heat,
protect against damage from harmful molecules called free radicals, and keep

their energy factories (mitochondria) working well. These systems stop harmful cycles without stopping growth.

Some animals can enter torpor or hibernation, lowering their metabolic rate and temperature to ride out famine or cold. Humans do not hibernate; however, controlled hypothermia saves brain tissue after ischemia, and researchers are continuing to cautiously investigate reversible human torpor efforts.[17] What they must ensure is to avoid unnecessary wear and tear, to invoke protective states when appropriate, and to acknowledge that species boundaries do not yield to will.

Finally, architecture and modularity represent foundational support strategies rather than direct defenses. Trees, corals, and colonial invertebrates are not long-lived because they maintain every cell in pristine condition; they are long-lived because failure does not spread. They use physical boundaries, replace parts locally, and build on inert supports. Our bodies are not trees, but our health systems, devices, and engineered tissues can borrow the same logic: Isolate damage, design for replacement, and treat redundancy as a feature, not a cost.

Myth and Precision: Words Matter

A few points are worth stating plainly. The "immortal jellyfish" does not float forever; it has a reset trick that sometimes works (and is beside the point if a fish eats it). Tardigrades are not superheroes in space; they survive insult in a paused state, then revive if rehydration comes in time. Hydra's negligible senescence is a statistical description of mortality risk in the lab, not a romantic claim that an individual hydra can live through human civilizations. And even the best-protected genomes still accumulate mutations. This precision in the concepts keeps expectations aligned with the evidence, making extrapolations less prone to wishful thinking.

From Field Guide to Work Plan: Translating Strategies Into Human Interventions

Now that we have learned from nature, we can consider implementing strategies into human interventions.

Near-Term Interventions

Initial efforts should focus on enhancing DNA protection and repair mechanisms. These could include standard health measures like smoking cessation, UV protection, and improved shielding in radiation therapy. Molecular enhancement, such as the use of small-molecule modulators of repair pathways, is also a possibility.

Proteostasis and autophagy reveal immediate potential. Experts have conducted experiments in animals with rapamycin and its analogs that show promise in extending lifespan.[18] In humans, these drugs are used for improving immune function. However, our focus should remain on identifying regimens that enhance function without compromising safety, considering risks like impaired wound healing and glucose homeostasis.

Mid-Term Interventions

Another practical implementation is through senescence modulation. The results in mouse models are positive when used for treating various diseases, encouraging ongoing trials, yet our translation to broader uses must be cautious and based on robust safety data.[19]

Metabolic stress management and protective states, such as therapeutic hypothermia, can be explored further, with careful consideration of pharmacologic torpor for spaceflight and taking into account species-specific considerations.

Other practices, like regeneration and partial reprogramming, demand strict precision. They hold the potential to restore function, yet the risks of erasing identity and triggering malignancy remain real. Studies and trials continue, and incremental steps include refining delivery, dosage, and surveillance mechanisms.

Moon Shot Interventions

Metabolic tuning, with an eye on systematic stress responses, may one day become central to making significant strides in longevity. Meanwhile, extracellular matrix and microenvironment engineering are emerging fields with substantial implications, though progress will likely be slow and tissue-specific.

Finally, I'd like us to focus on developing architecture and modular replacement strategies. This involves creating devices and grafts that can be swapped, ensuring that replacement is local and efficient. By mirroring nature's approach, we may prolong function even if core systems face challenges.

How to Tell Signal From Noise

One of the key themes of this book is that claims should be counted only when they alter hard outcomes. In longevity research, that means fewer hospitalizations, lower disease incidence, reduced mortality, and improved function in defined cohorts compared with matched controls, measured across years, not weeks.

As we engage with these scientific pursuits, we must also consider fundamental ethical principles such as autonomy, justice, and beneficence. The moral stakes must be considered alongside the scientific ones, ensuring that the allure of novelty does not overshadow caution. It also means accepting adverse events and publishing them honestly, as they also contribute to scientific progress.

Benchmarks like "epigenetic age clocks" are interesting, not definitive. In the previous chapter, we proposed "hype filters" for AI and digital continuation. So far, we have preregistered trials, independent replication, longitudinal registries, reimbursement codes that follow the evidence, and actuarial tables that adjust in response. Everything else, including new clocks, spectacular single-site case series, and press releases, belongs in "interesting but unproven."

Quantitative Anchors for the Curious

Numbers provide discipline and delight. Much of scientific progress is measured in quantified data. Below are some relevant numbers to illustrate what we have learned from science:

- Hydra replace epithelial cells quickly, on the order of days, and maintain that pace for years under culture, consistent with negligible senescence.[20]

- Planarian fragments regenerate into whole animals within days to weeks, depending on their size and species, guided by gradients and neoblast behavior.[21]

- Dehydrated tardigrades revived after days of space exposure,[22] and X-ray damage to cultured human cells dropped substantially when Dsup was expressed.[23]

- Massive *Porites* corals often extend linearly by about one to two centimeters per year, building centuries-scale archives of ocean chemistry in their skeletons.[24]

- Bristlecone pines add rings so slowly that the tree's cambium outlives entire civilizations.[25]

In terms of practical applications in humans, quantifiable outcomes such as a reduction in hospitalization days for patients could serve as a benchmark for metabolic stress interventions.

By anchoring our approach in specific numerical metrics, we can align with the standards of rigor established elsewhere. The point is not to fetishize numbers but to keep our imaginations moored to reality.

Equity and Measurement: Who Benefits, Who Does Not?

The goal of studying Earth's immortals is to develop human systems (clinical, legal, and cultural) that enable longer, healthier lives for more people without exacerbating inequality or creating new harms.

Data centers utilize modular, hot-swappable server racks to ensure continuous operation and facilitate easy upgrades. Similarly, we can consider modular tissue design. This approach enables localized repair and replacement without the need for a complete system shutdown, making it a durable and adaptable solution in healthcare. Implementing this approach would require measurement beyond the lab. Registries must track who receives regenerative grafts, senolytics, or reprogramming-adjacent interventions. Other important data includes patient outcomes categorized by ancestry, income, and geography.

Paying organizations should tie payments to specific results and demand long-term safety checks for coverage. Public reports should show access and problems by zip code. If there are gaps, these measures would allow for adjusting and reallocating funds. Otherwise, living longer will be available to only a few, not everyone.

To operationalize this vision of equity, we can implement a brief checklist after each proposed intervention, focusing on key metrics such as cost, accessibility, and data transparency. This checklist would need to include the following:

- evaluating financial barriers to access

- ensuring comprehensive geographical reach

- maintaining transparency in data collection and reporting methods

By making equity measurable, we can effectively turn these theories into successful interventions.

What These Lives Teach Builders

Engineers and clinicians think in stacks, constraints, and failure modes. Nature is a rich source of inspiration.

Hydra, tardigrades, and bristlecones are not miracles; they are systems with reasonable defaults. They defend DNA, keep proteins in line, replace parts, wall off damage, and route function around inert supports. The medical analogs are verifiable:

- Make repair and surveillance routine rather than heroic.

- Favor designs that tolerate local failure.

- Budget explicitly for energy, heat, and error rates.

- Keep humans in the loop, with independent audits of tools that claim to extend life.

The civic analogs are equally important:

- Rotate leadership and control rights to prevent power from ossifying.

- Train clinicians and regulators to update protocols in response to evidence rather than hype.

- Build redundancy into supply chains so that therapies do not vanish when a vendor fails.

If the tree's lesson is "Don't let small wounds faze you," the hospital's lesson is "Don't let a power glitch erase a life." These insights lead us to consider what frameworks are necessary. What concrete policy or design change can you implement tomorrow to make these lessons actionable in your context?

Bridges to What Comes Next

The organisms explored in this chapter also address themes that extend beyond biology, many of them igniting debates to come later in this book. Modular corals and compartmentalized trees preview debates about plural identity and copied selves. How much change a system can absorb and still count as "itself" is a preliminary question to start the debate.

Regeneration and partial reprogramming foreshadow neurotechnology's promise and peril. How much should we allow ourselves to reset, and who decides when reset becomes erasure?

Dormancy and energetic prudence open a discussion of cities and economies. What does stewardship look like when time horizons stretch and energy mismanagement compounds? The equity lens links to the psychological level, where we question infinite choice and the ethics of endless relationships.

Convergence, the meeting of AI, biotechnology, and neurotechnology that we made central in earlier chapters, will be necessary here as well. Pattern discovery, design, and verification can accelerate discovery; clinical craft and governance must then convert that acceleration into safe, shared benefit.

A Closing Without Romance

Hydra do not beat time; they adapt to it. Tardigrades do not conquer space; they survive by entering a state of hibernation when conditions are harsh. Bristlecone pines last by growing slowly and focusing on what matters.

We are not these organisms, but we can learn from them. If we want longer, healthier lives, we need to build systems at the cell, clinic, and community level that protect us, allow for repairs, measure progress honestly, and share the benefits fairly. These long-lived species are not our future, but they can teach us valuable lessons. However, there is more to analyze. Still within the biological realm, what potential consequences does longer human life bring to natural ecosystems?

Chapter 4

Ecosystem Dynamics

Immortality in a Finite Biosphere: The Problem Statement

Indefinite lifespans give people more time to care for the world but also more time for harm to build up. In the previous chapter, we explored nature to find cues that would allow human beings to expand life. Now, it is time to focus on what consequences this would bring to nature.

Scientists warn that surpassing critical boundaries, such as a 450 ppm CO_2 concentration, could lock in centuries of warming.[1] This poses a severe risk to societies, particularly those where natural resets are absent. In societies where people die, consumption and waste are limited by the lifespan of each person; belongings are passed on, infrastructure is replaced, and memories of damage fade. In a society where people live forever, there is no natural reset. Materials like buildings, electronics, and plastics can pile up for centuries unless they are designed to be reused or safely recycled.

On the other hand, indefinite lifespans will enlarge populations, leading to another set of challenges concerning resources. With more people, energy use and land development will continue to grow. Daily life changes and digital lifestyles will become more prevalent, as people will expect greater comfort over time. This implies more and more sophisticated devices and household facilities, creating a cycle of resource usage and waste accumulation.

The most significant risk is that robust systems, such as AI, robotics, and advanced bioengineering, can push too hard toward narrowly defined goals. For example, a farm focused only on short-term yield can fail if the climate changes; a power grid set up for average use can break down in an unexpected storm; and an ecological model based on limited data can miss a crucial interaction until it is too late.

However, immortality also offers potential benefits and trade-offs. A society with indefinite lifespans may benefit from accumulated wisdom, allowing for more informed decision-making and improvements in long-term planning. Certainly, this demands fluent and comprehensive communication across generations. Additionally, peer-to-peer education from experienced individuals could become a widespread asset, enhancing communal knowledge and driving responsible innovation.

An extended time horizon would also encourage investments in sustainability and resilience. Individuals who face the consequences of environmental decisions might be more committed to advancing effective solutions, because the impact will directly affect their lives; the risk and prejudices become more tangible.

This means that immortality is not inherently harmful to the environment. We need to shift our focus from just doing less harm each year to staying within safe limits for thousands of years and actively repairing past damage. To implement this change, I suggest we should do at least three things:

- Agree on the limits of Earth's systems.

- Translate those limits into clear budgets and rules that guide action.

- Establish technical and social habits that prioritize resilience, diversity, and the ability to reverse mistakes over merely increasing efficiency.

I dedicate the following sections to explaining each of these steps and connecting them to earlier themes. With these concepts and guidelines, you will have some clues to help you reflect on what you can do to be an active participant in this change.

A Safe Operating Space: Planetary Boundaries as Civic Accounting

The Earth has limited natural resources, many of which are scarce. Its planetary boundaries define specific physical limits within which humanity can safely operate. Crossing these limits significantly increases the chance of disrupting Earth's stable environmental systems. The framework used to define these limits originated from work by Rockström and colleagues and has been refined since then. It identifies nine processes, including climate change, biosphere integrity, freshwater use, land system change, and biogeochemical flows (nitrogen and phosphorus). Researchers set cutoff points for each of these stages, often expressed as ranges.[2]

Some boundaries, such as climate, are global in nature because, regardless of the region that particularly harms the environment, the consequences can't be restrained by any means. Other boundaries, like freshwater and nutrient loads, are regionalized, with local overuse often masked by global averages. Novel entities, such as synthetic chemicals, persistent pollutants such as PFAS, and potentially poorly regulated biological releases, add a new layer of risk that is harder to quantify but no less real.

We are flooded with alarming reports about the overuse of natural resources and the calamities it could bring. For immortals, data about these planetary boundaries would go beyond mere information and instead become an accounting system that needs to be interpreted in straightforward actions.

Let's explore some potential actions in which policymakers and relevant stakeholders on the global scene could engage:

- Translate global ceilings into allocable budgets per capita, sector, and place, with safety margins to account for uncertainty and ensure justice.

- Integrate "safe" with "just" by ensuring that meeting ecological ceilings does not come at the expense of basic human needs.

- Ensure access to clean water, nutritious food, clean energy, green space, and a healthy environment across communities.

- Create a public score, not just to spread information; updated data will keep people engaged in the conscious use of resources and aware of the importance of wise distribution.

Consider a drought-prone basin where water allocation must balance ecological ceilings and social foundations. In this region, water allocation is based on both environmental limits and human needs. During a dry spell, precise water budgets are calculated to sustain surrounding ecosystems while ensuring local communities retain access to essential water supplies. Farmers receive clear guidelines for water-efficient crops, and urban planning adapts to maximize water reuse and conservation. By involving community leaders in water management decisions, the region exemplifies how "safe and just" principles co-govern resources, maintaining both ecological health and social equity.

To achieve this, we can take carbon budgeting systems as an example. The city of Seattle provides a significant experience to look at, as it uses a Climate Action Plan to allocate and enforce carbon reduction targets across various sectors. Another example is the European Union's cap-and-trade system, which sets emission caps and allows for trading within those limits. These real-world models demonstrate how innovative governance and market mechanisms can be utilized to translate the planetary boundaries framework into actionable policies. The result is a dashboard that the public can read, that institutions must meet, and that courts and regulators can use to evaluate projects and policies.

From Diagnosis to Practice: A Public Stewardship Dashboard

A dashboard is only helpful if it measures the right things. While the categories may be familiar, what matters here is that the data is open, audited, and linked to actions. Some key variables and how they should be measured are listed below:

- Climate data should track yearly emissions and total budgets that align with global climate goals, broken down by sector, so that each area (such as electricity, industry, buildings, transport, and land use) is responsible for its share.

- The amount of healthy ecosystems should be measured by land use, the rate at which different types of land are being altered, and the quality of forests, not just the total area. These measures should support goals such as protecting 30% of land by 2030 and, where feasible, moving toward protecting half of key ecosystems.

- Freshwater data should indicate the volume of water flowing into each basin, the amount of groundwater used in comparison to the amount replenished, and whether water quality meets health standards.

- Nitrogen and phosphorus limits should be established for each region to guide farming and wastewater management plans.

The health of the biosphere should be tracked by examining trends in species, the connectivity of habitats, and whether ecosystems are providing essential services such as pollination, nutrient cycling, flood control, and supporting complex life.

Besides these familiar headlines, we can note a "novel entities" category to track the production and environmental fate of persistent chemicals, polymer additives, and other compounds that do not belong in food webs or groundwater. The practical outcome would be a phased schedule of bans for the worst actors, safe substitutes, and closed-loop controls for unavoidable uses.

The dashboard should also track the reuse of materials across the economy. It should measure the energy, water, and carbon used by computing, which is especially important in societies where digital life is a central aspect. Looking at power usage alone is not enough. We also need to consider the emissions from manufacturing hardware, measure the amount of water used and released, and ensure that any growth in computing aligns with energy and water limits.

None of these uses of technological devices is separate from medical efforts to ensure longevity. On the contrary, they are intertwined elements. Essential services, such as hospitals and water systems, should demonstrate their reliability and performance in emergency drills. As science and technology advance, our dependence on technological devices, which demand energy and other resources, increases. Therefore, it is necessary to measure both to ensure a wise allocation and use of resources, and also to understand how the equipment performs when something unusual happens.

All of this data should be available on a public platform, with clear methods and open audits. Here, I mean that information must be organized and published on free, easy-to-access platforms, and also that anybody should be able to read the dashboard and its data, not just specialists. Common people will have their lives directly affected by environmental issues, and the planetary ecosystem requires full commitment from all its rational inhabitants.

Institutions That Make Stewardship Stick

Dashboards without institutions to support and engage in direct action are decor. To convert numbers into norms, we need bodies with mandates and power.

Boundary Accounts Authority

The first institution to be created must be the Boundary Accounts Authority. Its key functions would be as follows:

- to maintain budgets derived from planetary boundaries

- to issue annual balance sheets that show our current standing

- to hold the authority to approve, deny, or condition projects based on their budgetary impacts

When a boundary is at risk, the Authority will be able to trigger corrective actions, such as increased fees, moratoria, or accelerated investment in offsetting and restoration. The Authority's dashboard should be flexible and rotate, while its methods must remain open. The Authority should be subject to an independent ombuds review to reduce the risk of overreach or corruption.

The Authority's remit would be to accept and address decisions taken by a participatory process involving diverse stakeholders. These stakeholders should include citizens' juries and local councils, whose insights and proposals would ensure shared rule-making and legitimacy.

This collaborative model echoes the success of commons governance, where integrated decision-making has proven effective. For instance, New Zealand's government created the Quota Management System in fisheries management, which is designed to sustainably manage fish stocks by setting limits on the amount of each species that can be caught annually.[3] The system allocates individual transferable quotas to fishermen or companies; these quotas represent a specific share of the total allowable catch for a particular fish species. The Quota Management System aims to prevent overfishing, promote conservation, and ensure the long-term viability of fish populations while supporting the fishing industry's economic interests. It allows for more efficient and responsible fishing practices through regulated catch limits and the ability to trade quotas; it also illustrates how legal frameworks can regulate resource use sustainably, allowing stock levels to replenish over time.

Similarly, California's cap-and-trade program effectively controls greenhouse gas emissions by setting a statewide limit and enabling businesses to trade allowances.[4] The program sets a statewide "cap" or limit on the total amount of greenhouse gases that can be emitted by regulated entities, such as power plants, industrial facilities, and fuel distributors. This cap is gradually lowered over time, which means the total allowable emissions decrease, pushing businesses to reduce their emissions accordingly. Under this system, the state issues a limited number of emission allowances, each representing the right to emit a specific amount of greenhouse gases (typically one metric ton of CO_2 equivalent). Businesses must hold enough allowances to cover their emissions. If they emit less than their allowances, they can sell or trade the surplus allowances to other businesses that need more, providing a financial incentive to reduce emissions. This example showcases how a well-designed system gains legitimacy and compliance through clear regulations and economic incentives.

Perpetual Conservation Trusts

Second, Perpetual Conservation Trusts would fund the protection, Indigenous guardianship, and restoration of lands and ecosystems in perpetuity. These lands would be a donation and, as such, these trusts wouldn't be driven by profit. Only the earnings could be spent under normal conditions, while the original capital invested could be used only in case of emergencies.

The revenue stream would be diversified and could entail the following:

- boundary fees on carbon, nutrients, and habitat conversion

- extended producer responsibility levies on high-throughput products

- a small tithe on the energy used by continuation services

The latter is directly related to the privileges of immortality and the obligation to engage in efforts to ensure biodiversity recovery.

Governance of these lands would be co-managed with Indigenous and local communities, because they have situational knowledge of how to keep the natural balance and strengthen local engagement and participation. This is aligned with ethical commitments and evidence that showcases how these communities often achieve superior conservation outcomes when their rights are secure.[5]

Shared Responsibilities

Producers are the third relevant stakeholder to consider. The level of responsibility here is as high as policymakers, local governments, and consumers. Producers in categories like electronics, batteries, textiles, and construction materials must be financially responsible for take-back, reuse, refurbishment, and safe recycling.

To support these processes, producers must adhere to detailed standards of repairability and disassembly. They could do this by creating a series of documents to ensure materials are efficiently used and disposed of. For instance, material and product passports would enable producers and controllers to track and repair items over their extended lifespans.

These documents would also regulate the use and treatment of toxic additives, which would preferably be phased out in favor of safe substitutes. Producers should also establish standards of life-cycle assessment that help set public buying rules. New product designs should reduce carbon emissions, conserve water, minimize pollution, and protect ecosystems, rather than focusing on a single factor.

An Ecological Central Bank

The basic point here is that resources are scarce and need to be allocated according to consensual criteria and attend to broad needs. The purpose of an Ecological Central Bank would be to translate ecological scarcity into price signals without pretending that price alone can manage plural values.

An Ecological Central Bank would set corridor-based fees payable by producers who emit carbon, release nitrogen, or convert habitat. When boundaries tighten, rates would rise, and the resulting funds would flow into the Perpetual Conservation Trusts (if these funds were taken into general budgets, they could be reallocated away from ecological purposes).

The Bank should also have an analytic role, modeling interactions between boundaries to avoid solving one problem by deepening another.

Safety and Reliability Regulator

Finally, another key institution would be the Safety and Reliability Regulator. It would treat water, power, waste, and continuation data centers as critical infrastructure. Among its functions, we can highlight the following:

- setting reliability service-level agreements

- requiring drills for failures

- mandating diverse supply chains and fail-safe modes

- publishing incident postmortems

The latter would provide evidence to support actions that are claimed as "smart must prove safe" or "green."

What Technology Can Do and What It Cannot

The contemporary world is framed by the accelerated advance of technology and the digital environment. Some of the most common applications are AI, sensing, and robotics.

These technologies bring several benefits that societies must harness as they reorganize to embrace sustainable longevity. For instance, AI and robotics are powerful in pattern recognition and optimization under constraints. Unlike human-driven tools, they are fallible when it comes to tacit judgment, legitimacy, and value negotiation. For all the rest, AI and robots, particularly after the development of agentic bots, can bring unprecedented benefits to societies and nature.

In ecology, they can detect anomalies promptly, with reduced use of resources and lower costs. For instance, AI can detect a fish stock's acoustic signature deviating from trends, a forest's thermal stress pattern shifting ahead of browning, or a wastewater nutrient spike upstream of a wetland.

Besides interpreting large amounts of data, AI and robotics are important tools for exploring possibilities beyond reality and anticipating the future. They can run counterfactual scenarios at scale and design portfolios that minimize risk in uncertain environments. They can also manage predictive maintenance and help prioritize restoration investments when budgets are limited.

Despite offering so many advantages, these technological devices have certain limitations. They cannot replace field knowledge, community trust, or the plural ends we choose. Let's consider a few realistic examples:

- An optimization that drives farms toward a single cultivar may look efficient until a pathogen arrives.

- An unexpected change in conditions might alter the decision-making process driven by the agentic robots.

- A water allocation algorithm that uses historical data can be misleading in a nonstationary climate.

These systems are more likely to work under stable patterns. When conditions are fluctuant, they may fail to draw accurate conclusions.

Let's imagine a government or producer that relies solely on a single model to control a system. It would be more vulnerable, and failure could easily escalate to the national scale if the model propagated a flawed assumption. To mitigate these risks, we could employ diverse model ensembles rather than a single engine. Additionally, we could utilize red-team ecological control systems, log and publish failures, and stage rollouts with real rollbacks rather than assuming success. Most importantly, we could constrain automation where reversibility is low, keeping humans in the loop for irreversible actions.

The human perspective wouldn't be replaced when it came to evaluating risks or handling randomness. This would ensure that every emerging factor was considered when allocating resources, navigating a crisis, or evaluating results, not just to ensure accuracy but also to preserve people's active participation. This model would give communities a voice in decisions that affect their landscapes. Open methods and data build trust, while privacy protections for community input protect dignity.

To effectively integrate technology with human oversight in policy or design contexts, consider these best practices:

- Diversify models and approaches to reduce reliance on any single technology.

- Involve communities directly in decision-making to ensure local needs and knowledge are represented.

- Ensure transparency in data and methods to build trust.

- Conduct regular audits and reviews of automated systems to detect and address failures quickly.

- Prioritize human oversight for decisions with significant, irreversible impacts.

- Protect privacy while leveraging community input.

Certainly, these actions need to be tailored to meet the unique characteristics of different communities. Each region in the world has a different starting point. For instance, some systems might need only to adapt audit processes to new amounts of data, while less accountable systems will need to create one from scratch.

Circular Materials and the End of "Away"

Circular materials are resources that are designed to be reused, recycled, or regenerated in a way that minimizes waste and environmental impact. Unlike linear materials, which follow a "take–make–dispose" pattern, circular materials are part of a closed-loop system where they continuously cycle through usage and recovery.

This approach promotes sustainability by reducing the need for new raw materials, conserving natural resources, and lowering pollution. Circular materials are commonly found in industries aiming for eco-friendly practices and sustainable production.

To avoid being overwhelmed by long-lasting products, we need to make circularity an integral part of the design from the start, rather than adding it as an afterthought. This means building products to last, making them easy to repair, and allowing for upgrades.

At present, many circular materials are used in industries and manufacturing:

- **recycled aluminum:** aluminum that is melted down and reshaped for new products, saving energy and resources

- **reclaimed wood:** wood salvaged from old buildings or furniture, reused for new construction or crafts

- **recycled plastics:** plastics that are processed and remolded into new items, like bottles, packaging, and textiles

- **bio-based plastics:** plastics made from renewable biological sources like corn starch or sugarcane, designed to be biodegradable or recyclable

- **recycled paper:** paper products made from postconsumer or postindustrial waste, reducing the need for virgin pulp

- **fabric made from recycled fibers:** textiles created from recycled cotton, polyester, or other materials for clothing and upholstery

- **glass cullet:** crushed recycled glass used to manufacture new glass products, saving raw materials and energy

However, this practice should not be limited to production; consumers must also develop this as a habit. Governments could distribute manuals with instructions on how to use standard parts and guidance on how to disassemble and repair items, allowing them to remain useful for many years. Governments and organizations could help by buying products that are repairable and come with more extended warranties.

Take, for example, a smartphone designed with a 20-year lifespan in mind. Each component is modular and can be easily swapped out as technology improves or if it becomes damaged. If the battery degrades, it can be replaced without requiring an entirely new device. The screen, made from durable materials, can be detached and updated independently.

Materials Passports

Materials passports, mentioned briefly above, would be documents with details of what something is made of, its use tracking, and its environmental footprint. These documents would be available to anyone, accessible via a simple scan. To continue with our smartphone example, it would provide users with information about the phone's composition, repair history, and recyclable elements.

This life-cycle story illustrates how design principles translate into everyday experience, extending the product's life and minimizing waste.

Extended Producer Responsibility

Extended producer responsibility (EPR) gives manufacturers skin in the game. It is an environmental policy approach where producers are given significant responsibility, such as financial and/or physical, for the treatment or disposal of postconsumer products.

The idea is to encourage manufacturers to design products that are easier to recycle, reuse, or dispose of in an environmentally friendly way. EPR shifts the responsibility from governments and taxpayers to producers, promoting waste reduction and sustainable product life-cycle management.

In this emerging scenario of an increasing population, take-back rates in the high 90s would be non-negotiable. Disposal would be the exception rather than the default. This approach would be costly, but one effective mechanism could be a "product return deposit."

This process would be similar to bottle deposit schemes, where producers charge an up-front fee refunded upon product return, encouraging consumers and businesses to participate in the recycling loop. Passports would carry records of a product's composition, repair history, and hazards, and would travel with items through resale and refurbishment. This simple label would empower consumers as auditors and make the process verified and transparent, covering carbon, toxicity, and biodiversity.

The issue of "novel entities," often overlooked, now demands serious attention. Persistent chemicals that do not belong in biological systems must be scheduled for elimination unless an essential use can be defined and controlled. Substitutes should be tested for both improved safety and improved performance. Microplastics should be addressed at the source, such as textile design, tire formulation, and wastewater filters, rather than treated as inevitable litter downriver. The measure of progress here is simple: We need to count tonnages of hazardous chemicals entering commerce, decline and recovery rates for critical materials, and an economy-wide circularity indicator.

Energy and Computing: Clean, Reliable, Audited

Immortal societies will be computationally intense. Digital continuation, clinical AI, and the background operations of cities all depend on power, increasing the pressure on natural resources. More power consumes more resources and, depending on the source, contributes to global heating.

The ecological answer is not to moralize against computing. Instead, the accurate approach is to clean and harden its supply chain and account for its externalities. Let's see greener, sustainable options, noting that many of these have already been implemented but not at the large scale that will be needed soon:

- Clean electricity must reach very high shares of production, and not only on annual averages.

- Wind and solar will do much of the work; they must be paired with storage, geothermal, advanced nuclear, and expanded transmission to cover nights, winters, and periods of low renewable energy.

The regulatory oversight discussed above would enforce drills for black-start capabilities and the islanding of critical loads, including water systems and continuation data centers.

Other measures can contribute to a rational use of energy, regardless of the source. A real-time "computing carbon counter" could be prominently displayed on public dashboards, translating digital energy consumption into visible metrics. This interface, featuring daily updates and clear visual indicators, would empower citizens to understand and engage with the ecological footprint of digital operations, turning abstract emissions data into tangible insights.

Power Usage Effectiveness

Power usage effectiveness (PUE) is a metric used to measure the energy efficiency of a data center or facility. PUE is calculated by dividing the total amount of energy used by a facility by the energy used specifically by its IT equipment. A PUE value of 1.0 indicates perfect efficiency, meaning all the energy is used by the IT equipment alone. The closer the PUE is to 1, the more energy-efficient the data center is. This metric helps organizations optimize energy consumption and reduce operational costs.

PUE provides relevant information but is insufficient, since computing's externalities also need to be disclosed and managed. For instance, there is embodied carbon in hardware, and computing leads to water withdrawals. Also, computing increases return temperatures; that is, there is an upstream mining impact.

When it is time to make decisions, the different stakeholders shouldn't focus only on chasing low prices; they should also reflect on regional energy and water budgets. Where thermal pollution poses a risk, cooling choices minimize harm. Meanwhile, air-side cooling or seawater systems should be preferred in suitable contexts, with monitoring in place to prevent local ecological damage.

We could also explicitly tie the benefits of immortality to biodiversity funding through a small continuation computing tithe. A fraction of the gross spend on energy and computing for continuation services would be allocated to Perpetual Conservation Trusts. That link would not be not a license to pollute. However, governments and societies will need high-energy, high-compute lifestyles, and they should finance the protection and restoration of living systems. To keep engagement and participation up, transparency is critical, so grants, outcomes, and failures should be public.

Biodiversity First: Protect, Connect, Restore, and Share Benefits

The Kunming–Montreal Global Biodiversity Framework established a global commitment to protect at least 30% of land and sea by 2030 and restore degraded ecosystems on a large scale.[6] Immortal societies can do more than meet a target; they can build institutions that maintain and deepen it across centuries.

Protection without representation is not protection. Rather, we must ensure ecological representativeness and management effectiveness. High-value biomes, such as tropical forests, peatlands, large carnivore ranges, and blue carbon ecosystems, are not simple patches on a map. They represent "Half-Earth" ambitions (the need to protect half of the planet's land and resources) and must be locally led and socially just.[7]

Protection

Protection only works if habitats connect, particularly in a dynamic environment. Recent research shows that large ecosystems are changing and species are becoming endangered as their habitats disappear.[8] To mitigate the damage, corridor networks across continents would enable species to migrate as climates shift. Also, road and rail retrofits with crossings would reduce mortality, and obsolete dams could be dismantled. This could be supported by utilizing permeability indices to guide investments to areas where small changes would unlock significant gains.

Restoration

Land restoration for diversity preservation refers to the process of rehabilitating and renewing degraded, damaged, or destroyed land to return it to a healthy, functional, and sustainable state. It involves activities that restore the land's natural ecosystem, improve soil quality, enhance biodiversity, and reinstate the land's ability to support plant and animal life. Land restoration can include reforestation, removal of pollutants, erosion control, soil improvement, and reintroduction of native species. The goal is to recover the ecological balance and productivity of the land for environmental, economic, and social benefits.

For example, China's Loess Plateau project consists of degraded lands that were restored to productivity and ecological function. The Netherlands' "Room for the River," which reconnected floodplains to lower flood risk and improve ecology, shows what is possible when investments and governance match natural time scales. The measure of success means survival and function after 5, 10, and 20 years.

As discussed earlier, Indigenous leadership is central. Where Indigenous land rights are secure, biodiversity tends to fare better. That is not magic; it is governance and relationships. Indigenous leadership enhances land preservation and biodiversity because Indigenous peoples often have deep, traditional knowledge and a strong cultural connection to their ancestral lands. Their long-term stewardship practices are based on sustainable use and respect for nature, which help maintain healthy ecosystems.

Indigenous communities typically manage land in ways that promote biodiversity, protect endangered species, and preserve natural resources. Their leadership also promotes community involvement and holistic approaches that integrate ecological, social, and spiritual values, leading to more effective and lasting conservation outcomes. Recognizing and supporting Indigenous leadership empowers these communities to protect their lands and biodiversity against external threats like deforestation, mining, and climate change.

Immortal societies should secure tenure, fund guardianship programs, recognize Indigenous and Community Conserved Areas, and implement benefit-sharing obligations under the Nagoya Protocol for genetic resources and traditional knowledge.

The Nagoya Protocol is an international agreement under the Convention on Biological Diversity that focuses on the fair and equitable sharing of benefits arising from the use of genetic resources, as well as the protection of traditional knowledge associated with those resources. Adopted in 2010 and entering into force in 2014, the Protocol aims to support the conservation and sustainable use of biodiversity; it does so by ensuring that countries and Indigenous peoples who provide genetic resources and traditional knowledge receive appropriate benefits when those resources and knowledge are accessed and utilized.

Key aspects of the Nagoya Protocol include the following:

- **Access and benefit-sharing (ABS):** The Protocol establishes a framework for obtaining prior informed consent and mutually agreed terms between users and providers of genetic resources. This ensures that benefits, whether monetary or nonmonetary, are shared fairly with the source countries or communities.

- **Protection of traditional knowledge:** The Protocol recognizes the rights of Indigenous peoples and local communities over their traditional knowledge related to genetic resources. It promotes the respect, preservation, and sustainable use of this knowledge and ensures that its holders receive equitable benefits when it is used.

- **Compliance and transparency:** Parties to the Protocol commit to establishing measures for monitoring and enforcing compliance with domestic ABS regulations, including checkpoints and information-sharing, to increase transparency in the supply chain of genetic resources.

- **Capacity building and technology transfer:** The Protocol encourages support for developing countries and communities to build capacity for ABS implementation.

Overall, the Nagoya Protocol strengthens global cooperation on biodiversity conservation. This supports the goals and well-being of immortal societies.

Oceans Protection

Oceans demand the same seriousness. Highly protected marine areas must expand, and fishing subsidies that drive overcapacity must end; rights-based fisheries with science-based caps can maintain livelihoods while allowing stocks to recover.

Blue carbon ecosystems, such as mangroves, seagrasses, and salt marshes, should be protected and restored not only for their carbon sequestration but also for their storm protection, nursery functions, and biodiversity. Satellites and automatic identification systems could monitor; prosecutions would require due process and international cooperation.

Working Lands and Waters: Produce Within Nature's Budgets

One of the most immediate consequences of an immortal society would be a larger demand for food. Thus, production of food, and specifically nutritious food, would be key. The ecological condition is that we must produce this food within climate, water, and nutrient budgets while supporting rural livelihoods.

This goal implies a series of conditions:

- climate-smart agriculture that reduces nitrogen surplus

- building soil carbon with cover crops

- reducing tillage to add habitat on farms with hedgerows and flower strips

- the use of precision application to minimize inputs

- livestock systems that reduce methane and manure emissions

- perennial systems that work in harmony with landscapes

- nutrient circularity (capturing phosphorus from wastewater, and composting urban organic waste rather than sending it to landfills)

- basin treaties that establish non-negotiable environmental flows

- water pricing that reflects scarcity

- basic human needs for households and small producers being the top priority

Fishery Resources

The same logic must be applied to fisheries. The allocation and usage of these resources also needs to be carefully planned and implemented. Science-based harvest control rules, gear restrictions to reduce bycatch, and enforced protected areas can sustain both stocks and communities.

Aquaculture is modernizing toward lower-impact species and methods, with waste capture and careful feed sourcing to minimize the displacement of impacts on other ecosystems. A real-life example of aquaculture is the farming of Atlantic salmon in Norway. Norway is one of the world's largest producers of farmed salmon, using aquaculture techniques to raise salmon in sea cages along its extensive coastline. The salmon are bred, fed, and harvested in controlled environments, which helps meet the global demand for seafood while reducing pressure on wild fish populations. This aquaculture practice supports local economies, provides employment, and contributes to food security. However, it also faces challenges such as managing water quality, disease control, and minimizing environmental impacts, which are addressed through regulations and improved sustainable farming methods.

Terraforming Earth: Carefully Named and Bounded

A new world needs new concepts and terms to name new realities, describe the emerging challenges, and imagine innovative solutions. We use the word "terraforming" with care. On Earth, the correct meaning is not planetary-scale geoengineering by default; it is ecological repair and design that restore function and increase resilience.

Nature-based solutions, such as reconnecting floodplains, retrofitting cities as "sponges" to absorb intense rainfall, restoring dunes and mangroves, and reconfiguring landscapes to reduce catastrophic fires, should come first for a reason: They work with processes that have evolved to handle variability and do not require constant, brittle control.

Engineered solutions should be allowed but controlled. The allowed projects would include changes to soil using tiny particles, devices that clean the air, and machines that manage land, all of which would need to be thoroughly checked. Specialists would need to look at several aspects: their energy use, the materials needed, how they might fail, and if they can be undone.

These solutions should be reviewed by independent experts and tested in steps, with the option to reverse them if needed. When materials that could cause big problems are needed, we must consider getting them from space to avoid harming Earth's environment, but only after the rules and technology are ready. Even then, we must have strong safety measures to prevent damage both on Earth and in space. The goal is not to reject technology completely but to prioritize fixes that can be undone and cause less harm.

Urbanization and Land Sparing

Increased populations impact directly on cities and unfold a series of problems. Nonetheless, societies can deploy potential solutions. Dense, high-amenity cities can reduce per capita footprint in the following ways:

- shortening travel distances

- lowering household energy use

- making shared services economical

If coupled with the intentional rewilding of de-urbanized territories, these strategies can spare land for nature at scales that matter. The spatial patterns, incentives, and lived experiences of such cities are discussed later in the book. The ecological claims are set out below:

- land-sparing targets should be explicit

- the people most affected by de-intensification should lead the planning

- rewilding should be measured in terms of ecological outcomes and human well-being, rather than in press releases

In a world of immortals, there is time to do this carefully; there is also time to let damage deepen if we fail to act.

Closing: Saving Resources to Survive and Thrive

Living longer should not mean making bigger mistakes. Instead, it should help us care for each other and the planet over the long term, aligning our goals with what the Earth can support. We should aim to leave behind more life than we take.

Tools like budgets, dashboards, laws, and practice drills turn hope into real, lasting progress. Suppose people who live forever want to be good stewards of the natural world and future generations. In that case, they need to focus on precise measurement, responsible care, and building institutions that remember what matters, even as time passes.

Throughout these chapters, we have explored nature: its richness, limitations, opportunities, and challenges. Now, it is time to turn to the protagonists of this new world: human beings. How will life change for people from each individual perception? There are so many aspects to ponder: philosophical, psychological, ethical, and emotional. Let's start our exploration with an open mind.

Part III: Identity in Flux—What It Means to Be Human When Humanity Is Redefinable

Chapter 5

Living With Your Copies

When Lena Came Home Three Times

The night when Lena introduced her second copy to the family, the roast grew cold. Seated around the table were three versions of her:

- Lena-Prime, sitting confidently at the head of the table, cheeks glowing from the oven's warmth

- L-2, a keen digital presence living inside a projector orb, eyes shaded by the stillness of someone who never sleeps

- L-3, a hybrid in a rented caretaker body, her hands steady after weeks of eldercare shifts

Opposite them, her mother gently stirred a glass of water, trying to hide her trembling, while Lena's teenage son flicked his gaze between the three: testing, measuring, choosing. In the doorway, the family's matriarch, a centuries-old digital grandmother full of charm and steel, spoke softly from a slim speaker: "Welcome home, all of you."

Not long ago, this would have been just a scene from a sci-fi story, but today, it is a portrait of family dinners after digital continuation has advanced. These common rituals bring together our values, our accounts, our promises, and our silent betrayals. When your life splits into several versions, these rituals become something more; they turn into structured systems. When your life becomes many, the pressing question of responsibility arises and grounds subsequent decisions. We must address emerging questions concerning identity: Who pays for groceries if one of you doesn't eat? Who goes to parent–teacher night? Who gets the house? The answers extend beyond family dynamics and reach broader societal notions of memory, power, and fairness.

The House With More Than One of You

After the singularity, we must consider three broad modes of persistence: biological, digital, and hybrid. The existence of these three versions of a person change cohabitation and intimate life. Let's identify each of them:

- the biological version, the one with the body, known as the "Lifer"

- the digital version existing as a computing model, known as the "Ghost"

- the hybrid that combines some elements of both realms, known as the "Shifter"

Each of them has a particular way of interacting with the other versions and with the surroundings. "Lifers" experience a constant, tangible connection to their physical form, while "Ghosts" sacrifice sensory feedback for rapid movement; "Shifters" navigate between both realms, instantly switching their focus from the dinner table to the digital network. Picture the sharp chill of cold metal pressed against Lena's skin as she transitions, the aroma of her mother's meal fading away as the quiet hum of servers fills her senses. It inspires contradictory feelings.

These three versions create a plural infrastructure built upon advances in brain mapping and simulation. The technology used includes nanometer-scale reconstructions and exascale substrates that surpass a mere theoretical experiment and instead make plural subjectivity a fact.[1]

The social outcomes echo earlier fractures in human history. In the past, human life has gone through deep changes that motivated communities to recalibrate their identity and rituals. For instance, in the late Middle Ages, France entered a crisis of fragmentation, and institutions reimagined the corpus mysticum to hold together what was splitting apart.[2] Similarly, with contemporary developments, the rise of digital twins in manufacturing and healthcare mirrors how entities are conceived beyond traditional physical boundaries.

This opens debates about the legal personhood of AI and highlights evolving perceptions of identity and the scope of rights in a digital age. Below, we will witness the transitional nature of identity and cohabitation and gain insights into how life will be in the immediate future.

Daily life will reflect these recalibrations. A digital grandmother might attend dinner through multiple avatars, speaking in several rooms at once, while a biological grandson still craves the heft of a hug and the brittleness of a laugh. "Shifters" will reconcile both tempos, and the person can be present in both the kitchen and a virtual office at the same time, combining intuition and the instantaneous.

As families change, rituals will evolve to adapt. Let's think about simple celebrations, like birthdays. Families will need to send dual invitations: physical cake at 6 p.m., virtual fireworks at 7 p.m. Memorials, on the other hand, will be part wake, part archive, with stories retold in a living room and memories replayed in a shared simulation. The new family calendars won't only save dates; they will also display who will be embodied, who will be remote, and who will be both.

These new manners will require technological devices. Nonetheless, the major concerns are related to ethics. We need to reconsider how to weigh presence, how to rank claims, and how to preserve dignity across differences.[3]

Besides these questions, there are structural issues that impact choices. Social division and economic shortages might be the most common. Immortal societies will create new places within cities. A key facility will be the Ascension District, a vibrant hub dotted with longevity clinics and augmented parks. Those with a certain income will be able to choose augmentation and digital continuation, and will move to this district to enjoy longevity. Others, instead, will gather in neighborhoods that deliberately maintain slower rhythms. Certainly, money won't be the only factor to influence these decisions; many will remain attached to traditional values. However, social class differences will limit people's choices and might shape their will.

If we think about social issues driven by inequality, we realize that differences in access will create new social stigmas and social exclusion. Imagine what life will be like in the Ascension District:[4]

- You're walking down the buzzing streets of an Ascension District, where neon signs flash advertisements like "Memory Enhancement Clinics Next Left" and the faint hum of drones delivering packages fills the air.

- A group of friends, their laughter trailing behind them, passes by, one saying to another, "I'll download the latest upgrade tonight."

- Offices learn new etiquette, including protocols for synchronizing memory downloads before meetings, quiet hours for augmented teammates, and reminders that "reading the room" now encompasses multiple rooms and substrates.

- Schools run dual curricula, including accelerated paths for students whose cognitive feeds can absorb a unit in minutes and project-based tracks that honor hands-on learning and organic pacing.

- Policies struggle to keep pace. The legal framework needs to rule the new situations while preserving the right to continue living according to traditional values and lifestyles, while discouraging discrimination against either status.

Beyond legal rules, societies will need to create new social bonds and ways to connect to build shared values and goals. In this world, living with your copies will be a family-sized version of a much bigger challenge: figuring out how different kinds of people can live together, even when their needs and abilities are not the same. And that challenge starts with money.

Shared Lives, Split Ledgers

Money clarifies. For households with multiple copies, the question will be less "Who earns?" than "Which version is on the hook for which obligations?"

Actuarial Curve

The lack of an actuarial mortality curve and mandatory retirement will also modify the labor market. Without a predictable exit, experienced workers may occupy rungs for centuries while new entrants crowd into lateral projects and founder roles.[5]

The existence of different copies of one person will allow parallel specializations. Thus, an executive could spawn a negotiation-optimized self and a crisis-operations self, each compounding skills through divergent experiences. The same person could hold more than one professional career through their various versions.

Adapted Organizations

Professional services firms that were once characterized by pyramidal structures will turn into guilds of polymaths. Companies will implement different strategies to organize their employees. For instance, they might opt for volunteer or program sabbaticals and "knowledge hibernations" to prevent tenure from becoming ossified power.[6] Since processes will be longer, reviews might focus on adaptability rather than on end goals.

HR Portals

HR portals, designed for individual users, will face challenges when "Lena" refers to three payroll profiles. Managers will need to adapt benefits and decouple them from identity to tie them to instantiation. For instance, health insurance for embodied versions might compute credits and redundancy insurance for "Ghosts" and hybrid riders for "Shifters."

Disability will become multidimensional, because it won't be related only to physical aspects. It might involve the loss of a body, the loss of a server farm, or the loss of a branch's continuity after a failed backup.

Unemployment insurance will evolve into "downtime cover" for employees whose roles are taken offline. Companies will seek to avoid role conflicts or comply with noncompete agreements. The lack of jurisdictional consensus on "instantiation insurance" will highlight a regulatory gap, and there will be emerging battles in the workplace. These issues can be addressed through the following proposed steps:

- Include model statutes that define and mandate insurance policies specific to each form and ensure coverage across all forms.

- Use regulatory sandboxes to trial these frameworks in controlled environments, facilitating the evolution of labor laws that recognize and accommodate the unique challenges posed by plural existence.

The firms that adapt best will be those that shift away from exclusive employment contracts, opting instead for scope-of-work agreements with specific branches or departments. They will use clear noncompete agreements to stop a negotiator from working with a competitor. They will also send work through smart contracts, which are digital agreements that track who did what.[7]

Household Finance

Families will build "branch ledgers" that allocate income and expenses across versions. Meanwhile, they will pool for shared goods such as housing, education, eldercare, and compute infrastructure.[8]

Taxes

Multiple versions of the same person open questions about tax obligations. Should the revenue from L-2's consulting in the virtual markets be treated as Lena's income, as the corporate income of a "Lena Collective," or as the income of a distinct legal person?

Without reform, households will face double taxation or, worse, tax invisibility that might lead to public backlash. To manage this, adaptive governance suggests starting with transparency:[9]

- Register branches.

- Apportion income by instantiation.

- At the policy level, decide whether to tax by head, by identity, or by household.

Wealth Concentration

None of this touches the deeper engine: wealth concentration. In an immortal economy, wealth will accumulate uninterrupted in the absence of taxes. The traditional pressure valve, which is taxation on inheritance at death, fails when there is no one to inherit. Copying exacerbates "legacy lock" because the founder persists, and so do their branches, each reinforcing the initial concentration.

Take, for instance, the hypothetical Johnson family, who lack access to longevity technologies. Wealth concentrates in the hands of those who can afford or navigate the copying process, while families like the Johnsons find themselves perpetually on the margins of prosperity.

This explains why societies turn to renewal mechanisms, such as those listed below:[10]

- sunset clauses on control rights

- charter expirations that force companies to reapply or reconstitute

- identity-based caps on voting power so that one origin cannot multiply its influence through various branches

- cyclical resource reallocations that prevent permanent dynasties while preserving incentives to create

The principle is mainly about circulation. Renewal is the condition of shared prosperity in a world where people have the power to decide how long they will live.

Early Adopters and Resisters Under One Roof

Economic redesign collides with culture at the front door. Lena's household spans the spectrum: a biological mother who treasures the heft of memory, a hybrid daughter who treats identity as a portfolio, and a digital grandmother who has outlived half her friends and most of her glories. While this may become the new normality, it will come with challenges.

In offices, the augmented staff will brainstorm at speeds that unmodified colleagues struggle to follow. The latter will also fail to reach promotion as opportunities tilt toward the enhanced. Office jokes will develop a new social language to refer to "analog brains" and "living antiques."

In schools, parallel curricula will include adapted content that promises fairness, but this could also lead to stratification, separating "Shifters" from unmodified students. In neighborhoods, safety nets and social rituals will diverge: weekly potlucks in one district, multisensory salons in another.[11]

Legal Framework

New charters must aim to consecrate bodily autonomy, which entails the right both to remain unmodified and to modify. These charters must also preserve peace and human dignity with nondiscrimination rules that apply to hiring, housing, and public services.

Civic institutions should adopt accessibility standards in both directions: captioned virtual events for embodied participants, and tactile, quiet spaces for those who cannot or will not process high-intensity augmented feeds.

Governments will need to commission task forces on resource allocation, asking openly what justice looks like when some can choose immortality and others refuse or lack access. Additionally, cultural institutions should do their part, with museums curating "living exhibits" of pre-enhancement crafts, digital temples rising beside old churches, and poets writing of the fragile and the infinite.

No matter how many changes institutions and governments implement, conflict will remain. It always has. But policy can blunt its sharpest edges and teach households, by example, how to honor different modes of being. People and the collective building of culture will do the rest.

The Will and the Double

Let's learn more about Lena's family. Lena's father was a founder with an iron memory and a soft spot for rule-bending and complicated things. Years earlier, he created a digital copy, then a hybrid. Finally, he added a cluster of role-specific variants to run various aspects of his firm. There was a rainmaker for investor relations, a monk for R&D, and a tactician for compliance.

He promised Lena's mother that these were tools, not replacements, and that his will would be drafted before the branching and leave everything to "my spouse and my child, per stirpes." Yet, when his biological body finally failed, the copies immediately complicated matters, arguing that "he" had not died in any meaningful sense. The digital father claimed continuity, the hybrid father asserted control of the operating shares, and Lena-Prime argued that "per stirpes" meant the human line, not the copy line. Her mother, in turn, asserted her right to an elective share.

Most legal systems were built to track unitary persons, not a cluster of versions of the self. Still, there is no need to invent everything from scratch. Existing tools can be repurposed if we name the problems properly.

Identity and Capacity

Courts must decide whether a copy is a continuation or a descendant. Psychological continuity theory recognizes each copy as a valid successor if the two entities share similar memories and values. However, it cannot identify a single true heir when multiple copies exist, since each one represents a separate psychological continuation rather than the exact same person.[12]

Animalist views prioritize the organism, denying identity to copies and, by extension, denying them succession rights based on sameness.[13]

Practically, family courts and probate judges will need a registry that establishes branch identifiers, determining immutable tags that tie each instantiation to an origin event and a divergence time. Other documents, such as contracts, wills, and insurance policies, can specify obligations and benefits by branch ID rather than by a name that now covers many.

This also carries ethical concerns, particularly the privacy costs we might be prepared to pay to guarantee accountability. We must also reflect on potential surveillance abuse in the pursuit of rigorous identification.

The dynamics among jurisdictions make the scenario even more complex. Different legal systems, such as civil and common law, may approach the establishment and enforcement of branch registries in diverse ways. In civil law jurisdictions with prevalent codification and detailed statutes, regulatory frameworks may emphasize legal tools for defining and managing branch identities. Legal systems that rely on past court decisions change slowly, influencing how branch registries work through specific court cases. This variety can help start conversations about global rules, encouraging more discussion on how to align legal systems while respecting different cultures and methods.

Death and Survivorship

Jurisdictions can define the "civil death" of the original instantiation upon declared branching, and can also recognize the continued personhood of copies for all other purposes. This would allow life insurance to pay out and probate to proceed without erasing the beneficiaries' rights.

Alternatively, lawmakers could consider continuity trusts. When someone starts a new branch of the family, their assets would go into a trust, which would benefit the spouse, any children, and each branch of the family. Rules would decide how much control each branch has, giving more weight to branches created earlier or those that have cared for family members.

This control could decrease over time to stop one branch from staying in charge forever. Trusted managers could be chosen from outside the family to avoid conflicts. The trust might also require regular, verified records showing each branch's contributions to family care and business management.[14]

Intestacy, Elective Share, and Marital Regimes

In community property systems, branching complicates the line between marital and separate property. Does the value created by a copy post-branch accrue to the marital estate or to that branch alone? Default rules could treat pre-branch assets as marital and post-branch gains as branch property, subject to compensation for the use of marital infrastructure (e.g., housing, computing).

Elective share statutes are designed to prevent disinheritance and should be extended to protect spouses from dilution by proliferating copies. A spouse might choose to remain married to one designated branch, convert to a consortium marriage that includes all branches, or dissolve the marriage with equitable distribution. The least unjust option relates to procedures:[15]

- Require notice before branching.

- Implement cooling-off periods.

- Provide mandatory disclosures to the spouse.

- Enforce clear lines for support and custody.

Corporate Control and Voting

A founder could replicate to multiply their votes and entrench their control. Corporate charters can avoid this by tying voting rights to the original identity rather than headcount, or by capping the aggregate votes held by branches of a single identity.

Besides voting rights, sunset provisions can phase down control over time, with reauthorization requirements every few decades forcing boards to justify their continuation under renewed terms.[16] Instruments should also define "bad branch" conduct, such as self-dealing and asset tunneling, and should empower boards to quarantine or deactivate a branch's governance rights upon finding a breach, subject to independent review.[17]

Digital Assets and Memory Estates

Even without copying, digital remains bring headaches for heirs. With copies, the problem multiplies. Who owns the raw memory archives after a branch is deleted? Who controls the social media feeds and the feelings that tie families together? This can be improved, but a starting idea is that the spouse or children manage memorialization while the branches manage operational archives, subject to privacy constraints. Any memory-sharing technology should be subject to consent rules clarifying who may access, edit, and deploy memories in therapy, litigation, or art.[18]

Marriage and Family, Rewritten

Marriage vows written for two weaken with three. The default law of many jurisdictions treats bigamy as void; copying scrambles that rule. A default rule might treat branching by a married person as creating a multiparty marital estate, with the spouse electing to do one of the following:

- Remain married to one designated branch.

- Convert to a consortium marriage with all branches.

- Dissolve with equitable distribution.

The spouse's decision should trigger mandatory mediation, not as moral theater but as operational planning. For instance, it should establish calendars for physical presence, budget divisions, and what to do on holidays. If we seek success, we can't expect simplicity, and we certainly must not avoid complexity. The basis is to name who is responsible for what, and then enforce it.

Children and Care

Co-parenting across branches raises practical questions, such as pickup schedules, medical consent, and school meetings. These are about daily organization, but other questions tap into complex parenting matters: Who has tie-breaking authority?

Family law already privileges continuity of care. Courts could extend that norm to branch networks, measuring caregiving. The child's voice should matter. In Lena's case, for instance, her son told the judge that Lena was "the one who shows up." That expression of desire should be enough, but much of family law neglects common sense and privileges procedure.

Relationships

More versions of you might mean more friends and relationships but will also create unknown challenges. Friends will ask which "you" they are talking to. Partners will wonder who they love or who loves them back. Each copy is the individual in a way: Each has the first-person glow of "I am me." But even if we assume each copy is authentic, this doesn't mean equal obligation for all of them. Promises should track the branch that made them, unless renegotiated;

consent cannot be imported from a sister branch; and forgiveness should be a matter for one branch at a time.[19]

Memory-sharing might increase empathy, but on the other hand, it destabilizes evidence. If Lena-Prime uploads a childhood memory to L-2, does L-2 "remember" the abuse to share in therapy or serve as testimony? Ethically, consent should track both transfer and use, and the law should treat shared memories as derived records (probative but not dispositive), because there are many risks of editing and context loss.[20] Courts will need to train experts to distinguish between implanted recall and lived experience, not to diminish the former's moral weight but to prevent manipulation in litigation.

Identity Coherence

Without an endpoint ahead, identity might vanish. Humans give meaning to their existence by noticing beginnings, middles, and ends. Branching replaces the end with a mesh of variants. So, how would people find meaning in their lives?

Periodic "identity audits" would help. These audits imply guided check-ins to align values across branches, prune overlaps, and reassign roles if needed. New cohorts of therapists will be trained in immortality stress and will counsel clients through survivor's guilt, temporal disorientation, and the burden of being first to set the rules. They will rely on the same tools employed until now (narrative therapy, life mapping, strategic memory pruning adapted to branch networks), preserving continuity and an integrated self.[21]

A Case Unfolding

At the first probate hearing, Lena's lawyer introduced the problem to the court: "We cannot ask metaphysics to do the work of equity," she said. "Treat the copies as descendants, not continuations."

The proposal was as follows:

- Recognize "civil death" of the origin for estate purposes.

- Pour the assets into a continuity trust.

- Distribute an elective share to the spouse.

- Allocate stipends and education benefits to each branch.

- Assign voting control to the branch most engaged in caregiving and firm operations.

- Set a sunset in 20 years and reauthorization by independent trustees.

In exchange, the branches would stop insisting on "Who is the real husband?" and accept a "one identity, one vote" rule in civic matters to avoid multiplying political power. The judge signed off on the proposal.

The corporate counsel wired the governance on a ledger. They gave a tag to each branch and established a scope for every right, enforcing the fiduciary duty. The lawyer warned that the smart contracts would not solve grief. Still, they would prevent the worst abuses: asset tunneling by a rogue copy, vote-packing at shareholder meetings, and the endless "but I am him" of a founder who could not stop becoming.[22] They assigned votes per origin identity and inserted a branch quarantine clause. According to this clause, there would be an automatic suspension of governance rights upon an audit-verified breach, with a review by an independent panel.

Home life was still complicated. Holidays, which always test patience, made the differences stand out even more. On Thanksgiving, Lena's versions pulled the family in different directions. L-2 wanted to celebrate in a virtual version of the old family farm, evoking memories of childhood gatherings. Meanwhile, L-3 felt attached to the hospital where she underwent her shift, finding purpose in care and a sense of urgency. Lena-Prime, immersed in the warmth of her mother's kitchen, wished to savor her grandmother's recipes. Their solution was a mix: together in person at midday, online in the evening, and some quiet time alone at night. The family kept track of it all, and so did the child, learning that family is what you choose to keep doing together.

At work, the team overcame the problems. Lena-Prime's employer complained that L-2's consulting competed with the firm's strategy practice. A noncompete clause drafted years earlier did not mention copies. Arbitration decided a pragmatic split:

- L-2 could consult in sectors where the firm had no footprint.

- Lena-Prime could continue to lead in-house engagements.

- Both would use the registry to prevent double-billing and conflicts.

The union was at first reluctant to accept branches, but eventually it organized them. Solidarity across branches was the only way to prevent employers from pitting workers against each other.

There was a mistake, too. In family therapy, L-3 shared a copied memory of a teenage fight between Lena and her mother: a slammed door and a cutting remark. She meant to bridge understanding. Instead, her mother felt invaded. "You weren't there," she said. The therapist found a new way to reverse this harm. They wrote a rule: no sharing of intimate memories without the consent of all participants. When it comes to family, love can fix what isn't written in theory or rule manuals.

Suppose a jurisdiction defined "civil death" upon branching for probate. In that case, a poorly timed backup could determine which branch inherited operational control. This absurdity can be avoided only if laws and families agree to attach rights to branch IDs and presence rather than to the fiction of seamless continuity.[23]

The Technical Scaffolding Underneath

The move from unitary to plural selves rests on imaging and computation that capture the brain's structure and its plasticity. Existing methods help us understand this.

Functional magnetic resonance imaging (fMRI) maps functional flows in living brains, especially when combined with targeted molecular tools. Also, electron microscopy and light-sheet imaging enable the resolution of synapses at the nanoscale. High-throughput reconstruction transforms petabytes into workable models.[24] Specialists also rely on simulations that require hardware that can run massively parallel processes with reliability. This hardware includes neuromorphic chips and quantum-accelerated clusters.

We will need to develop special software that can host both structure and change: memories forming, habits dissolving, selves growing. However, there is another problem related to storage. A full scan at nanometer resolution generates exabytes, so specialists will need to compress and address selectively with ethical consequences. Every system relying on power is vulnerable; a consciousness might die without a steady energy supply, just as it would from physical harm.

These constraints will shape household decisions. A daily backup would ensure minimal loss if a branch failed but would also freeze memory at the moment of backup, raising questions about which version is considered "current" in the event of a dispute. Families will need to negotiate backup windows with the intimacy once reserved for trying for a child, while family budgets will need to include computing expenses alongside groceries and rent.

Designing Fairness

Living with your copies is about design. Here are our choices for the design of households, firms, and politics.

Registries and Rails

Implement branch registries that fix identity genealogies at the moment of divergence and require their use for contracts, benefits, and court filings. Tie the registry to verifiable, privacy-preserving infrastructure with enough transparency to assign liability and allocate rights, as well as enough obscurity to protect dignity. Add cryptographic signatures to specific branch IDs to prevent the laundering of obligations through a fog of selves.[25]

Privacy safeguards can be further enhanced by employing zero-knowledge proofs, which allow verification of data or actions without revealing the underlying information. Additionally, establishing independent oversight bodies can help ensure accountability and address surveillance risks.

Consider the challenges: If a branch were to conceal a crime, how might this registry uncover it? Suppose a branch attempts to falsify financial statements by creating fake identities. The cryptographic signatures tie each action to an authenticated branch ID, allowing for easy tracing of discrepancies back to their source and triggering alerts for independent review.

Sunset and Churn

Build sunset clauses into the control rights of companies, trusts, and data to counter legacy lock and open lanes for new entrants. Require corporate charters to reauthorize periodically, and allow public interest boards to compel renewal on different terms. Erase super-voting rights over time.[26]

Presence-Weighted Obligations

Tie family obligations such as support, custody, and decision authority to measurable presence and caregiving, not to metaphysical claims of authenticity. Keep a schedule of care: hours at school, nights in the hospital, and holidays delivered. Use them as tiebreakers when love can't settle a tie.[27]

Civic Parity

Prevent political distortion by capping civic power to "one identity, one vote," with internal branch governance choosing the voting representative by lot or rotation. Without such caps, a wealthy origin could flood a district with branches to swing elections. Equality before the law relies on parity of political representation, despite the plural identity.[28]

Beyond these, a handful of practical policies ease pressure points:

- Antidiscrimination laws must cover enhancement status and branch status, protecting both resisters and adopters.[29]

- Education funding should flow to high-speed curricula for augmented learners as well as to programs that preserve crafts and community.

- Healthcare must treat compute outages as health crises for digital citizens and provide coverage for hybrids and biologicals.

- Labor law should presume that branches of one origin cannot be treated as substitute labor for anti-union tactics.

- Courts should develop specialist dockets for identity cases, drawing on interdisciplinary expertise in neuroscience, contract, family law, and ethics.[30]

However, the political economy obstacles cannot be ignored, as entrenched interests might awaken resistance. Corporations that benefit from current power dynamics, traditionalists who fear cultural erosion, and even segments of the population concerned about the implications of technological advancements may oppose such initiatives.

To alleviate this resistance, policy adoption strategies could include providing incentives such as tax credits for companies investing in inclusive technologies and subsidies for educational institutions that implement dual curricula. Meanwhile, campaigns might raise public awareness to encourage cultural acceptance. These campaigns could highlight the benefits of plural existence, while dialogue platforms could bridge gaps between traditional and augmented communities. These steps could enhance policy uptake and ensure a smoother transition toward an inclusive future.

A Household Checklist

A family bringing a new copy home can learn from corporate onboarding and ancient rites. Here's a checklist that will help:

- Register the branch, obtain a branch ID, and update all existing contracts (employment, leases, loans) to avoid ambiguity about who owes what.[31]

- Amend wills and beneficiary designations to reflect the branch network; consider a continuity trust with sunset governance.[32]

- Draft a household operating agreement that allocates expenses, divides labor, and sets rules for computing budgets, backups, and redundancy.

- Establish memory-sharing protocols: what can be shared, with whom, and for what purposes. Include consent from other people in shared memories.[33]

- Clarify romantic commitments and boundaries across branches; apply "promises follow presence" unless renegotiated; memorialize agreements in writing.

- Set co-parenting schedules and tie-breaking mechanisms, measure and log caregiving, and give children a voice in age-appropriate ways.

- Audit noncompete and confidentiality obligations across employment contracts to prevent an accidental breach by a sister branch.[34]

- Schedule identity audits, quarterly at first, to align values, prune overlaps, and reassign roles; include mental health support for burnout and meaning drift.[35]

- Decide on civic participation: which branch votes, serves on juries, and runs for office. Rotate to distribute the civic load.[36]

- Build rituals that honor each mode and celebrate contributions beyond income: care, patience, presence.[37]

Why the Rituals Matter

In social life, policy restrains the worst, and rituals nurture the rest. Rituals ensure a sense of community and create belonging. That's why they matter.

In Lena's house, they established small ceremonies. On Sundays, each version named one thing she admired in a sister branch. On the first of the month, they reviewed the agreements with gratitude to reflect on what might go unnoticed: the long wait at the DMV, the bug fixed at 3 a.m., the awkward apology made to a friend. Twice a year, they pruned: L-2 wound down a side project that was taking up family time; L-3 passed the eldercare shift to a neighborhood cooperative; Lena-Prime gave up trying to cook for everyone and started hosting dinners just for "Lifers."

These habits didn't eliminate conflict, but they created space for it to occur. Jealousy still came up when L-2's work was praised but she didn't help at home. It still hurt when someone at a party asked, "Which one of you are you?" Through these rituals, the family shifted its focus from questions of who got what to visions of how they might flourish together.

As discussed earlier, philosophy has sketched this ground for centuries. Psychological continuity explains why multiple versions can each claim authenticity, while no one has exclusive title to the original's identity. Animalists still believe personhood is linked to the organism.

People will use subjective self-awareness, affirming "I am me" in every branch, but this is not a valid argument in court. The Ship of Theseus tells its quiet joke: Replace planks one by one or all at once, but either way, you will still need to decide what to call the vessel and how to dock it among others without collision.[38]

These philosophical questions guide the drawing of policies and laws. Psychological continuity supports the idea of creating identity registries, which can provide legal foundations for handling branching identities in governance. Animalism's focus on the organism suggests limits to other versions' agency. The Ship of Theseus metaphor highlights the need for laws and policies that can adapt to a flux of identities.

Once we accept that identity is plural after copying, the question changes from "Who is the real Lena?" to "What do we owe one another, given that each of us is Lena in some vital sense and yet not the others?" That question is not a puzzle to be solved; it is a practice to be repeated.

What Policy Can and Cannot Do

It is tempting to imagine that a perfect registry, a clever trust, or a crisp charter could neutralize all harms. It cannot. It can, however, establish baselines to address harms and overcome obstacles. The combination of instruments such as registries and rails, sunset, and churn is not anti-technology; it is pro-community. This approach frames and regulates the possibility of plural lives and then asks how to keep the commons intact when the temptation to dominate multiplies.

No matter the efforts, many people might still resist branching and remain unmodified. They must be protected from cultural condescension and economic exclusion. The right to remain finite is as sacred as the right to extend. The better vision is a mixed society where pace, mode, and number vary and where the law treats each of us as a citizen with equal claims to respect and resources.[39]

Closing: The Last Dinner in This Chapter

After the hearing, the family went home. L-2 quietly downloaded the day's records. L-3 washed the roasting pan and hummed. Lena-Prime stood at the sink, feeling the warmth of the water, and realized something: Living with your copies isn't about picking which one is real. It's about creating rules that respect each version's reality and make love feel genuine for everyone.

There will be more dinners. The roast will get cold again. Someone will say the wrong thing. A child will pick one of you, then change their mind. Laws will struggle to keep up. But if we're careful, if we build systems that remember, agreements that forgive, and traditions that respect both old and new, we can make a life together. The real work isn't about living forever. It's about being a good family, even when we're different. The promise isn't to end death but to make belonging bigger.

Even though these claims are the basis of social peace and human dignity, this chapter has opened many questions that will emerge with a new world. In the following pages, we will explore these debates in depth, not to find absolute answers, but to explore possibilities. We are creating a future: a brand-new, better world.

Chapter 6

The Three Humanities

A Taxonomy for Divergent Futures

This chapter maps three divergent futures and the safeguards that will enable their coexistence. While these futures might seem metaphysical, this book focuses on the processes and frameworks that will shape them.

As AI, biotechnology, and neurotechnology converge, our collective abilities will sharply transform. Institutions can only acquire responsibility when they are safe, verifiable, and shared fairly among stakeholders. Consequently, as people adjust, their identities will not converge but rather diverge into distinct paths. These futures will manifest in three distinct approaches to being human: some will believe in embodied life, using technology to support rather than replace it; others will prioritize patterns, speed, and omnipresence, emphasizing digital and augmented existence; and a third group will learn to navigate and translate between both worlds.

Throughout this exploration, we will revisit key ideas introduced in earlier chapters. We cannot exclude the following from our reasoning: the imperative for verifiable digital lives, an understanding of consent as an ongoing process, the need for transparent data provenance, the integration of fairness as a design principle, and the recognition of energy and reliability as fundamental ethical considerations. Indeed, without the necessary governance tools, identity could easily drift into unfair and inequitable systems.

Biological Purists: Embodiment First

For biological advocates, being human is not an abstract proposition. It is a practice. It is breath after rain and the way a hand remembers the weight of a cast-iron pan. It involves the rhythms of circadian light and dark, the body's reactions when determining whether a room is safe or not, and the social knowledge we build from reading faces and bodies in real time.

Imagine, for example, a visit to your garden: You step onto the cool, damp earth, feel the texture of the soil between your fingers as tomatoes ripen, and inhale the blend of grass and fresh blooms. In these moments, life is a sensory experience.

Those who advocate for biology accept medicine. They prioritize repair and prevention, and they want these services to be reimbursed and thoroughly audited. They disagree with the idea that people need to be improved to fit into a system. They also reject that being modern means making permanent changes to the brain without a clear reason, and that the benefits are greater than the risks.

They welcome regenerative medicine when it helps avoid organ failure or long-term disability. No one objects to treatments that keep a kidney functioning or reduce tissue damage under close supervision. However, when the topic shifts to changes that cannot be reversed, such as permanent sensory changes or neural upgrades that can't be undone, they prefer options that allow reversal of the changes, ongoing consent, and time to adjust before making permanent alterations. This approach recognizes that the body teaches us, and that some of what makes life meaningful comes from facing challenges and limits, not from removing them all with technology.

Research has gathered evidence from nature exposure and multisensory experience that supports biological advocates' arguments. Studies link time spent in nature to reductions in rumination, improvements in mood, and even immune modulation. These effects are created by more than just what we see. They come from the mix of smell, touch, and the way light moves through trees in a way that our eyes like to take in.[1]

Other studies reveal that a good user experience depends on psychologically satisfying tools, whether digital or physical. The most meaningful lives give people room to choose how to engage, to get better at something over time, and to share both the practice and the result with others.[2]

Then, biology advocates employ the same basic health tools as everyone else. They focus on maintaining a good metabolism and sleep, incorporating useful treatments and medications when necessary, repairing organs when they fail, and monitoring health to prevent emergencies.

What Makes the Difference

What sets biological purists apart is what they choose not to accept. They do not see enhancements as moral duties. They suggest making reversibility the standard in medical guidelines and treating consent as something that can be renewed, like a license. For them, ongoing consent is a cornerstone of moral agency, reinforcing the human capacity to make informed, autonomous decisions over time.

They require clear energy limits and plans for what happens if a device fails. For them, a device that stops working under stress is not reliable care. They propose to track access by location and income, and they expect budgets to change if gaps grow.

Their ethics are practical. If a treatment changes how someone feels or thinks, there should be clear ways to undo it, regular breaks from aids to keep choices real, and ways to return to normal for those who want to. At the policy level, they suggest measures that include establishing regulatory standards to dictate reversibility in medical practices and mandating ongoing consent protocols. Furthermore, they believe funding mechanisms should be directed toward research and development of reversible medical treatments and technologies. Robust audit requirements, then, would ensure compliance with these standards, providing transparency and accountability aligned with these values.

Those who support these arguments are not hostile to digital continuation, but they refuse to grant metaphysical victories and prefer to live their lives as organisms. If, in the future, a digital agent presumed to be Aunt Reva passes the verification regimes (behavior under novelty, autobiographical recall with context, trait stability, relational validation, and transparent provenance), the biological purist community might invite Aunt Reva back to Sunday dinner. But this doesn't mean they don't believe something is missing from the original embodied being. In the meantime, they build their lives with the tools at hand: hands, gardens, networks of care, and the kind of attention that sees quickly when something living is "off."

Digital Consciousness: Continuity on Silicon, Under Constraints

It is one thing to talk about "mind uploading" on a stage, but a very different thing to convince a hospital, a court, and a family to act as if a digital agent is the same party who consented, promised, and parented when they were biological. I am insisting on verification regimes beyond philosophical arguments because they effectively reduce harm, fraud, and self-deception.

To make these regimes operational rather than aspirational, we can follow a repeatable checklist that can be established. Key steps in this checklist are as follows:

1. Preregistration to ensure initial alignment

2. Comprehensive auditing for transparency

3. Familial sign-off to maintain trust

4. An apparent willingness to abandon processes when tests indicate failures

Furthermore, institutions must clearly define the boundaries of use. For example, a continuity agent that supports a daughter through probate may not be authorized to run a company. Similarly, an agent that closes a promise may not be authorized to open a new line of credit. As evidence and governance improve, the contexts for these uses can broaden, but they should not begin with the broadest possible claim.

When people discuss digital life, they often make grand claims, such as "perfect memory" or "instant analysis." It is better to be specific. Digital memory can be organized and linked to shared archives, enabling models to identify patterns that people might overlook. Still, memory is stored in physical systems that can fail, and retrieving information depends on agreed-upon rules and permissions.

The idea of "perfect retention" is just a slogan. What matters is whether someone can ask a personal question and get a meaningful answer, including the correct details or even an honest "I don't know" when needed.

"Near-instant" is a goal, but the real challenge is to design systems that are both reliable and fair rather than making impossible promises. Fast analysis is possible when the hardware is close by, the models are well-configured, the data is clean, and the tasks are clear. However, these services rely on data centers and network speed, are subject to energy limitations, and can become slow under heavy usage.

New Sensory Experiences

If memory and analysis can be calibrated, we can expect technology to address the question of sensation. Virtual worlds already deliver a multisensory experience. In future chapters, we will explore design options. Digital life will rely heavily on those tools, and many digital individuals will continue to create or design embodied platforms; example uses might include time spent with children, or forms of work and art that feel more engaging with weight and breath.

Digital persons can do things biological persons cannot: branch, compress, and run processes at speeds that make new modes of collaboration possible. Conversely, biological persons can do things digital persons cannot without borrowing a body. Some of those differences will remain even as simulation quality improves. That is not a sign of failure. On the contrary, the most interesting communities will be mixed, and both groups will learn to connect and interact in healthy ways.

The most challenging aspect of digital governance is governance itself. Compute power is a critical infrastructure that dictates how many processes a person can run, how immersive an environment can feel, and how rapidly an individual can acquire new knowledge. Without clear rules and a foundational level of access, societies will inevitably develop a compute aristocracy that, disturbingly, mirrors old wealth but with more sophisticated branding.

For instance, the top 1% could command 50 times more processing power, highlighting how swiftly such hierarchies might emerge. However, we can learn from other past experiences. The necessary mitigations are similar to those employed in other public utilities: establishing compute commons to ensure a baseline level of access for all citizens, maintaining market transparency, implementing anti-hoarding regulations, and sharing surge capacity mechanisms.

For example, laws could establish a minimum compute allocation for educational institutions and public services, ensuring equal access to computational resources. Anti-hoarding regulations could prevent the monopolization of compute resources by setting caps on the amount any single entity can control, similar to spectrum allocation in telecommunications. Furthermore, institutions could regulate surge capacity mechanisms to ensure that during high-demand periods, compute resources are distributed fairly and efficiently, much like how the electricity grid is managed.

It is essential to account for carbon footprints with the same meticulousness applied to financial audits. Reliability audits are equally important, as we noted in the last chapter. Continuity that falters during a power outage is not continuity, just as continuity that loses its identity during a failover is inadequate.

Then, we need to consider engineering for redundancy and graceful degradation. Alongside these interventions, the legal framework must evolve in tandem. Chapter 5 introduced tools such as branch registries and continuity trusts to ensure that promises are linked to specific instances and prevent the uncontrolled multiplication of property and control rights. The same instruments must support digital civic infrastructure while enforcing parity protections, such as "one identity, one civic vote," even when an individual has numerous manifestations.

It is easy to imagine that these changes will happen quickly, but in reality, progress will be slower and more careful. Initially, continuity agents will be trusted only in limited situations, such as closing a promise, clarifying intentions, or assisting with technical or artistic work. In these situations, we can assume the facts are clear, or that we are reuniting families under protective rules.

These use cases will make it easier to identify problems and minimize risks. As these early uses succeed and audits improve, the range of uses can grow. Meanwhile, we should watch for clear standards from regulators and courts, truly independent audits, reports from families, and public data.

Hybrids: Bridges That Learn Both Speeds

Hybrids are not hard to imagine because many people already live this way. They utilize devices for enhanced focus, collaborate using shared dashboards, rely on AI assistants for specific tasks, and transition between physical and digital spaces throughout the day. In this chapter, we look at what happens when these tools become more reliable, closer to us, and better managed. We pay special attention to areas that require clear rules, including consent, safety, cybersecurity, labor, and fairness.

In a clinical setting, a hybrid life begins with devices that have multiyear stability and have earned their place through demonstrated patient benefit. Let's think, for example, about a person with chronic pain who functions without the fog of opioids because a stimulator breaks a loop that the body could not stop by itself.

In this case, success relies on raw bitrate; more generally, this approach relies far more on materials that the immune system tolerates. It depends on surgical protocols that reduce infection risk, on rehabilitation norms that teach people how to live with a device without letting it swallow their life, and on consent processes. The latter needs to describe what happens if the device fails at 3:00 a.m. or is hacked at 3:00 p.m.

The best clinical programs in this area already discuss patch transparency, device life-cycle plans, third-party audits, and red-team exercises that aim to break what is being deployed so it can be fixed before harm occurs. They also talk about the right to stop.

Reality Beyond the Lab Walls

Outside the clinic, hybrid life will require the establishment of norms. The most apparent risk is exploitation: the sense that someone who can keep a parallel thread running should keep that thread running for an employer's benefit at all hours. Healthy organizations will not let that happen. They will set quiet hours and decouple promotion from augmentation.

For enhanced organizational cultures, business will train managers not to infer "lack of grit" from lack of tools, and they will write interview norms that do not let the augmented run circles around the unaugmented in rituals. Healthy organizations will also recognize the invisible labor that hybrids often perform as interpreters between embodied and digital teams, and will compensate them accordingly. Without these norms, hybrid life would be a treadmill that burns people out and deepens resentments between groups. With them, it becomes an asset that can help a city run, a hospital learn, and a family manage a crisis.

It is essential to reiterate a key point from earlier: We should not treat new technologies, such as quantum computing, as panaceas. If a hybrid person uses specialized hardware for a specific task, it is a practical choice, not a shift in what it means to be human. Durable systems rely on well-tested practices, though depending on experimental technology increases vulnerability. As innovation advances, our priority should be designing systems that are both rational and dependable.

Cross-Platform Communication and Emerging Hierarchies

In the years ahead, one of the most important social skills will be translation. People who live mostly in bodies and those who spend much of their time in digital spaces communicate at different speeds and rhythms. Even when everyone wants the same outcome, this difference can cause misunderstandings.

This is just a practical issue, not a personal one, and it can be managed with clear rules. For example, in a meeting, the chair can ask digital participants to slow down and match the pace of human conversation. They can also ask those in the room to explain their assumptions and provide context that digital team members might miss, like who is speaking up and who is staying quiet. Hybrid colleagues, who move between both worlds, often make the best facilitators. The key is to recognize this work and create tools to support it.

Besides the subject component, tools matter. In virtual spaces, where many people will spend much of their time, presence design can make the difference between flailing and trust. Systems that allow an avatar to render facial signals with fidelity improve understanding; systems that render those signals too perfectly risk manipulation.

Consent controls for emotional telemetry should be at the frontline, too. The literature on VR presence and multisensory design is far enough along to give designers checklists for what helps and what harms.[3]

In physical spaces, the analog of avatar design is hospitality. We can create rooms where those who come with high-bandwidth streams in their minds can rest, alongside rooms where those who are digital by default can be present without interruption.

Power will follow resources unless communities take proactive measures to prevent it. Compute and network priority are not abstract. They shape who can precompute options and see farther down a decision tree. Without guidance, those powers will yield the same path-dependent advantages that people have exploited for centuries. Having equal resource time during decision-making can be helpful.

Therefore, we can use audit trails and fair scheduling rules. Embodiment produces its unique biases: Height, voice, and charisma have long advantaged some bodies over others. However, digital fluency can do the same. The recognition and rotation of modes and roles reduces those biases, and the law should do the rest. In Chapter 5, we explored the many aspects that laws and policies should consider to ensure rights are commensurate with responsibilities.

Equity remains the most challenging part. This can be achieved via compute commons, universal broadband, AI assistance for accessibility, and public reporting with budget triggers when gaps widen.

Parity protections will matter too. It must be illegal to discriminate by mode in hiring, housing, education, and civil services. Regulations must guide the development of habits and customs that accommodate both directions. Thus, companies, social organizations, and stores will need to hold virtual events for people with disabilities and provide quiet, tactile rooms and embodied hospitality for digital visitors. Targeted funding could support initiatives that bridge gaps, ensuring all areas receive equitable access to resources. These mechanisms will be crucial in holding institutions accountable and guiding the necessary improvements.

Integrating the Rails: How the Three Paths Fit Into a Single Polity

By this point, you are familiar with the arguments of previous chapters and will recognize the essential principles of digital identity verification and user consent that anchor this discussion. These include methods to verify and track digital activities, ensuring that consent can be easily renewed and changes can be easily reversed. If all three paths (embodiment, digital continuation, and hybrids) share these supports, then living together becomes something we can design, not just something that happens to us.

Let's consider a city zoning meeting where hybrid facilitators play a critical role. As the meeting unfolds, stakeholders from various backgrounds (embodied, hybrid, and digital) gather to discuss a new development project. The chair, aware of the potential for conflict, sets equal compute windows for digital participants while encouraging embodied members to share their insights verbally. Hybrid facilitators stand ready to bridge any communication gaps, translating technical terms into everyday language and ensuring all voices are heard. By the meeting's end, the participants have built consensus, enabled by clear protocols and mutual respect that encouraged collaboration.

In a city's climate office, a meeting that brings together gardeners and fishers, grid managers and water engineers, and a modeling team can be run without losing anyone's voice. In a household that contains a biological parent, a hybrid caregiver, and a digital grandparent, a branch registry can tie promises to specific instantiations, while a presence-weighted ledger can record care. The family can protect the embodied parent's right to a quiet hour and the digital grandparent's right to a stable connection. They can also mark holidays in both modes and speak to one another with respect across the gap created by the speed difference.

To make coexistence work, we need to avoid simple solutions that only *seem* fair. It is not fair to require everyone to use enhancements to be equal, just as it is not fair to avoid learning new tools simply because they are unfamiliar. We should not overwhelm elections with multiple versions of one person. In mixed groups, it is not fair to expect digital people to slow down without also asking those in the room to share context and explain their thinking. Most of these issues are practical problems that can be solved with clear rules, training, and flexible resources.

Setting Expectations: Signals to Watch

If you finish this chapter wanting to take action, you do not need a big, dramatic project. Instead, you need to know what to look for and what to ask from your institutions:

- A policymaker in healthcare might consider framing oversight as a series of questions: Are guidelines for elective enhancements prioritizing the "reversible first" approach? Is there a system in place for renewing consent? Are there public dashboards available that display access by area? Do insurers offer comprehensive long-term care coverage?

- In legal and administrative frameworks, someone might inquire whether established and enforced rules exist for digital continuity. Are audits conducted independently, and how commonly are continuity trusts employed in estate planning?

- In the workplace, you might want to develop criteria to assess the fairness of policies that affect both augmented and unaugmented staff. Are quiet hours respected across all departments? Do employment norms take into consideration different degrees of human augmentation?

- For municipal governance, teams need to ensure that shared computing resources are available and that systems for monitoring energy use and reliability are selected for their robustness in the event of failure. Oversight tools that allow the public to hold these systems accountable should be implemented.

- In educational environments, teachers and counselors should start by verifying if students are being taught to work efficiently across different speeds and modes. They need to ensure that educational tools are designed to make consent and pacing explicit and apparent.

We are all engaged in the task of designing better guidance, oversight, legislation, and programs for these futures. Each of us brings knowledge, expertise, and personal experiences to enrich the debate and outline collective solutions.

Closing: Coexistence by Design Rather Than Drift

The three humanities described in this chapter will not spontaneously inter-weave with the others. Communities will have to decide whether choice remains a reality and whether people who live at different paces can learn to trust one another. The rest of this book applies the same rails we have addressed at the individual and subjective level to a broader scope: to cities and economics, to governance without generational turnover, and to the varieties of experience people might choose when time is no longer the limiter of a life.

Keep the questions close: Who makes the decision? Under what rules? With what data? At whose energy cost? With what off-ramps? If you can answer those questions in your household, at your clinic, in your city, and in your workplace, then you will already be living in the future this chapter describes.

Part IV: The Architecture of Forever—How Society Rebuilds When Time Stops Mattering

Chapter 7

Population Plateau

The Great Stabilization: What a Plateau Is and Is Not

A population plateau does not mean everything stops; nor does it mean that something might decrease. Instead, it is an extended period, often lasting decades, during which the total population remains relatively stable but the composition of the population continues to evolve. This distinction is essential. In classic demography, a "stationary" population occurs when the number of births and deaths is equal, vital rates remain constant, and the age structure remains unchanged. An immortal society cannot fit this definition because the age mix constantly changes. People who would have died from aging survive, groups of people accumulate, and the population pyramid transforms into a column and, then, a top-heavy pyramid.

The easiest way to understand this is to picture the population as a flow. The population comprises various age groups. Births add people, while deaths remove them. If advances in longevity significantly reduce deaths among older people, fewer individuals will leave the population at those ages.

Suppose birth rates remain below replacement levels for an extended period, as is already happening in many places. In that case, fewer people are added because there are fewer parents and they have fewer children. During the transition, more older people survive, so the total population continues to rise even though births remain low. Once lower death rates and lower birth rates are spread across all ages, and other causes of death are more evenly distributed, population growth slows and levels off. This is the plateau I refer to. Reaching it takes time because age structures change slowly.

It helps to use simple numbers to make this more straightforward, even if we are not making exact predictions. Picture a world in the middle of the century with 10 billion people, and let's consider the following trends:

- If deaths from aging drop sharply, annual deaths could fall from about 90 million to 40 million.

- If annual births are between 60 and 70 million due to low fertility and fewer parents, population growth would drop to around 20–30 million per year, or 0.2–0.3%.

Over time, the number of older people would increase, fewer births would occur, and growth would continue to slow toward zero. Death would not disappear, since people would still fall victim to accidents, violence, infections, and rare events. But the old patterns of dying would become much less critical. The plateau would come slowly, and when it did, it would hide ongoing changes: the population would be getting older, but people would be staying healthier for longer.

It is essential to recognize what we do not know. Digital continuation may lower birth rates for some individuals by meeting their desire for legacy or occupying time that could have been spent on parenting. Other factors could push in the opposite direction. For instance, longer healthy lives make parenting over many years feel safer, while shared households could spread out the workload and cultures that plan for the long term might make late and intentional parenting more common.

To illustrate these dynamics, consider Maria, a 68-year-old grandmother who has chosen to dedicate her postretirement years to sharing her life experiences digitally with her extended family rather than actively participating in child-rearing. Meanwhile, in another scenario, James and Lisa, both in their mid-50s, have recently decided to become parents for the first time, taking advantage of their continued vitality and the supportive structure of a multi-generational living arrangement.

These examples highlight how evolving family choices can impact demographic trends. So, we should expect differences across regions and groups. The best approach is to treat these as ideas to watch, not as truths set in stone.

A City in the Plateau

Imagine a coastal metropolis in the 2070s. Its headcount has remained essentially unchanged over the past 15 years. What has changed is who makes up that headcount. The share of residents over 65 has risen steadily, yet the streets don't show any sign of decay. Instead, the air is alive with the chatter of vibrant interactions and the gentle hum of electric vehicles gliding by. The scent of freshly brewed coffee and baked goods wafts through the air as older adults walk, cycle, teach, run clinics, lead labs, and sit on boards after undergoing periodic recertification of their practice.

Median ages in university classrooms have spread as lifelong learners cycle in and out alongside first-time students. In lofts that used to be offices, multi-generational households share kitchens, studios, and childcare rooms, with one apartment reserved for a digital grandmother who joins dinner through a presence orb. The rhythms of the city remain cheerful, with the laughter of children playing in reimagined parks and the beats of distant music from cultural festivals. This is a futuristic scenario, but we can take a glance at it by looking at Singapore's innovative urban planning or Tokyo's integration of elderly-friendly public spaces.

The capital budget at city hall would appear unfamiliar to a planner from the 2020s: more than half of the infrastructure dollars are allocated to maintenance, retrofits, and state-of-good-repair, while fewer funds are earmarked for expansion. The planning department's performance dashboard is public and audited. The new headline metrics are "retrofit rate," "share of homes meeting thermal and air-quality standards," "tree canopy coverage by neighborhood," "take-back rate for electronics," and "reliability drills passed by hospitals and data centers.'

This picture is not meant to idealize the future but to bring some optimism to mitigate anxiety about reaching a plateau. When the population stops growing, the focus shifts from having more to improving quality, longevity, and fairness. What becomes valuable is careful maintenance, strong institutions, reliable clean energy, and space for renewal so cultures stay flexible.

Why Planning Metrics Must Change

Let's start by analyzing the relationship between gross domestic product (GDP) and population growth. Until now, certain key metrics and concepts have been used. Here's an explanation of these metrics:

- **GDP growth rate:** This measures how much the GDP of a country changes over a specific period, usually annually. It indicates the overall economic performance. When compared with population growth, it helps us understand if the economy is expanding faster or slower than the population.

- **Population growth rate:** This is the rate at which the number of individuals in a population increases in a given time period, often expressed as a percentage per year. It helps analyze how fast the population is growing relative to economic growth.

- **GDP per capita:** This is the GDP divided by the total population. It provides an average economic output per person and is a crucial metric to assess living standards and economic well-being. Changes in GDP per capita reflect whether an increase in population is accompanied by economic growth.

- **GDP per capita growth rate:** This measures the rate at which GDP per capita changes over time. It helps evaluate whether individual prosperity is improving or declining, considering both economic growth and population changes.

- **Dependency ratio:** This ratio compares the number of dependents (people younger than 15 or older than 64) to the working-age population. It influences GDP per capita because a high dependency ratio can mean fewer workers supporting more dependents, affecting economic productivity.

- **Labor force participation rate:** This shows the proportion of the working-age population that is employed or actively seeking employment. It affects GDP because higher participation can lead to higher production and economic growth.

- **Demographic dividend:** This concept refers to the economic growth potential that can result from shifts in a population's age structure, particularly when the working-age population is larger compared to dependents. It directly ties population dynamics to economic output.

- **Total factor productivity (TFP):** While not directly a demographic metric, TFP measures the efficiency of all inputs in the production process. It can help explain changes in GDP beyond just labor inputs affected by population growth.

In a plateau, these metrics become less valuable as a policy north star. What you need to know is excluded from these original metrics. Instead, we need to measure the following:

- whether the places where people live and work are healthy and resilient

- if material loops are closing

- whether the assets you've already built are being kept up

- if benefits are being shared

- if the energy and compute that undergird digital life are clean and robust

The data must be delivered on dashboards that report the right quantities and are linked to budgets, standards, and drills.

Cities and Housing

For the built environment, the headline quantities are a city's retrofit rate and the share of infrastructure spending devoted to maintenance.

In a mature city, you want a retrofit rate in the low single digits of total floor area each year. This is sufficient to decarbonize and adapt the legacy stock in a reasonable time frame, aligning with emission reduction targets and ensuring continuity with emissions timelines. A higher percentage, such as 10%, could lead to economic and logistical strains, potentially disrupting building use and access. A consistent majority of capital dollars should be allocated to maintaining assets in good repair.

In the new cities, investors and brokers will want to see vacant properties transformed for active use, highlighted by a decrease in long-standing vacancies in central areas, a rise in converting offices into residential units, and improved vehicle circulation. This last point includes easing parking requirements and replacing car lanes with bus routes, cycling paths, and tree-shaded sidewalks. This doesn't mean cars will be banned, but more convenient and safer alternatives to nearby destinations will make driving an uncommon choice.

Planners can draw inspiration from cities that have successfully implemented similar strategies. For instance, in Freiburg, Germany, the city has transformed its public transport network and reduced car dependency, demonstrating the viability of mode shifts. Freiburg has implemented a comprehensive approach to urban mobility that prioritizes public transport, cycling, and walking over private car use. The city has invested heavily in expanding and improving its public transportation system, including an efficient and well-integrated tram and bus network. This network provides frequent, reliable, and affordable service, making it a convenient choice for residents and visitors.

In parallel, Freiburg has developed an extensive cycling infrastructure, including dedicated bike lanes, bike parking facilities, and traffic-calming measures that make cycling safe and attractive. The city has also enhanced pedestrian areas, creating vibrant, walkable neighborhoods that encourage people to leave their cars behind for short trips.

To further discourage car use, Freiburg has implemented policies such as limited parking availability, higher parking fees, and car-free zones, which together help reduce traffic congestion and pollution. The city's urban design promotes mixed-use development, where homes, workplaces, shops, and schools are close to each other, further supporting walking and cycling.

Similarly, Melbourne, Australia, has pioneered office-to-residential conversions, revitalizing former business districts. As some traditional business districts in Melbourne experienced shifts in demand (partly due to changes in economic activities, evolving work patterns, and, more recently, the rise of remote work), the city faced the challenge of underutilized office buildings and declining foot traffic. To address this, Melbourne introduced policies and initiatives encouraging the conversion of vacant or underused office spaces into residential units.

These office-to-residential conversions have revitalized former business districts by introducing new residential populations into areas that were once primarily commercial. By repurposing existing buildings, the city has made efficient use of its urban infrastructure and reduced the need for new land development, supporting sustainable urban growth.

The influx of residents has brought vibrancy, economic activity, and a sense of community to these areas beyond traditional nine-to-five business hours. New demand for local services, shops, restaurants, and cultural venues has helped rejuvenate neighborhoods, making them attractive places to live and work.

These examples provide valuable precedents and data for cities considering a shift toward a maintenance-focused infrastructure development strategy.

Population and GDP

As we have expressed, equity is a major concern. We must be careful when addressing dashboards since averages mask inequality. GDP measures the total economic output of a country, but it has limitations in revealing how wealth and income are distributed among the population:

- **Aggregate measure:** GDP is a sum of all goods and services produced, so it reflects overall economic activity but does not show who benefits from this wealth. A high GDP could coexist with significant income inequality.

- **Ignoring distribution:** GDP does not account for how wealth is shared across different social groups, regions, or individuals. A few wealthy individuals or corporations could control most of the economic output, while many others might have very little.

- **Overlooking poverty and living standards:** GDP growth might increase overall wealth, but it can happen alongside persistent poverty or stagnant living conditions for large segments of the population.

- **No insight into access:** GDP does not measure access to essential services like healthcare, education, and housing, which are critical for equitable well-being.

- **Excluding informal economy:** In many countries, the informal economy is large and may not be fully captured by GDP data, obscuring the economic realities of lower-income or marginalized groups.

- **Environmental and social costs:** GDP growth might come with environmental degradation or social disruptions that worsen inequality but are not deducted from GDP figures.

- **Income vs. wealth:** GDP measures income flow but does not reflect wealth accumulation and disparities, which are key dimensions of inequality.

Because of these limitations, complementary metrics like the Gini coefficient, poverty rates, the Human Development Index (HDI), and measures of wealth distribution are essential to understand the true extent of inequity within an economy.

For equity, dashboards must be disaggregated by specific groups. Housing quality, green access, heat exposure, air quality, broadband, healthspan, and time-to-service after failures should all be visible by neighborhood and income, as Cattaneo and colleagues urge when they model development along the urban–rural continuum.

Energy Consumption

Measuring energy consumption and population growth involves using specific metrics that provide insights into usage patterns, trends, and demographic changes. Here are the key metrics for each one.

Energy consumption metrics are as follows:

- **Total energy consumption:** This measures the total amount of energy used by a country, region, or sector over a given period, typically expressed in joules, kilowatt-hours (kWh), or tons of oil equivalent (TOE).

- **Energy consumption per capita:** The total energy consumption divided by the population size; it indicates the average energy use per person.

- **Energy intensity:** This metric shows the amount of energy consumed per unit of GDP, reflecting the efficiency of energy use in economic activity.

- **Renewable energy share:** The proportion of total energy consumption that comes from renewable sources like solar, wind, hydro, or biomass.

- **Fossil fuel consumption:** The share and amount of energy derived from coal, oil, and natural gas.

- **Final energy consumption by sector:** The energy used in sectors such as residential, industrial, transportation, and agriculture.

- **Energy efficiency indicators:** Metrics that track improvements in how energy is used, such as energy saved per unit of output or service.

Population growth metrics must consider the below:

- **Population growth rate:** The annual percentage increase in the population, accounting for births, deaths, and migration.

- **Crude birth rate:** The number of live births per 1,000 people per year.

- **Crude death rate:** The number of deaths per 1,000 people per year.

- **Net migration rate:** The difference between immigrants and emigrants per 1,000 people.

- **Fertility rate:** The average number of children a woman is expected to have during her reproductive years.

- **Population density:** The number of people per unit area, indicating how crowded a place is.

- **Age structure:** The distribution of the population across different age groups; important for predicting future growth and economic impacts.

- **Urbanization rate:** The percentage of the population living in urban areas; often linked to energy demand patterns.

Together, these metrics help analyze how population dynamics influence energy demand and consumption patterns, and how efficiently societies use energy relative to their population size and economic output.

In the context of a plateau, the numbers must pierce the marketing and consider more specific and practical aspects. Data centers should disclose their renewable energy share, water withdrawals, and return temperatures, as well as the embodied carbon in expansions and the results of black-start and islanding drills. Hospitals, water systems, and data centers should be treated as life-critical infrastructure, prioritizing reliability and being subjected to drills overseen by an independent regulator. These service expansions should be contingent upon the procurement of clean power and storage that is commensurate with load growth.[1]

The actual test of a good metric is whether it forces a decision. Below are some interesting guidelines to consider:

- A maintenance share under 50% in a mature city should block expansion proposals.

- A take-back rate of less than 95% in a covered category should trigger corrective fees.

- A data center that fails its reliability drills should have its license revoked until it passes the required tests.

- A rewilding plan that does not secure the tenure and consent of Indigenous and local communities should not proceed.

These enforcement mechanisms resonate with Elinor Ostrom's principles of effective governance, which refer to a set of design principles for managing common-pool resources effectively and sustainably. These principles emerged from her pioneering research on how communities self-organize to manage shared resources like fisheries, forests, irrigation systems, and grazing lands without external enforcement. Her work challenged the traditional notion that common resources are inevitably overexploited (the "tragedy of the commons").

Here are Ostrom's eight core principles for successful management of common-pool resources:[2]

- **Clearly defined boundaries:** The resource system and the individuals or households with rights to use the resource must be clearly defined. Knowing who has access and what the resource boundaries are prevents free-riding and conflicts.

- **Congruence between rules and local conditions:** The rules governing resource use should fit local needs, conditions, and cultural practices. Rules that are tailored to the specific context are more likely to be accepted and followed.

- **Collective-choice arrangements:** Most of the individuals affected by the rules can participate in modifying and designing them. Inclusive decision-making encourages buy-in and better compliance.

- **Monitoring:** Monitoring of resource use and users is conducted by members of the community or accountable individuals. Monitoring ensures rules are followed and resource status is tracked.

- **Graduated sanctions:** Punishments for rule violators are graduated, starting with mild sanctions and increasing for repeated or serious offenses. This encourages compliance without being overly harsh.

- **Conflict-resolution mechanisms:** Accessible, low-cost methods exist to resolve conflicts among users or between users and officials. Effective dispute resolution maintains cooperation and trust.

- **Minimal recognition of rights to organize:** The community's right to self-organize and manage the resource is recognized by external authorities. Support from higher-level institutions legitimizes community governance.

- **Nested enterprises (for larger systems):** For resources that are part of larger systems, governance is organized in multiple layers (local, regional, national), with coordination and communication among them. This helps manage complexity and scale.

Regulation Based on Principles

Ostrom's principles highlight the importance of local knowledge, participation, adaptability, and legitimacy in resource management. They provide a framework to design institutions that can sustainably govern shared resources without top-down control or privatization.

For instance, graduated sanctions and local monitoring serve as proven community governance tools that can align with metric-based levers, thereby reinforcing trust in these governance frameworks. To embed these enforcement mechanisms effectively into city policy, planners could incorporate them into municipal ordinances or building codes, ensuring that the criteria for metrics are clearly outlined and subject to regular audits and public reporting.

Regulatory bodies may also consider developing specific permits and compliance checks that serve as checkpoints, allowing these metrics to be assessed and triggering necessary actions when thresholds are breached.

Social Metrics Adapted to the New Scenario

Creating social metrics to evaluate Ostrom's principles in real cases can help assess how well communities manage common-pool resources according to her framework. Here are suggested social metrics aligned with each principle:

- **Clearly defined boundaries:**

  ○ Metric: Percentage of community members who can accurately identify the resource boundaries and eligible users

  ○ Metric: Existence of formal or informal agreements defining user rights and resource limits

- **Congruence between rules and local conditions:**

  - Metric: Degree of alignment between rules and local cultural practices, measured through surveys or interviews

  - Metric: Frequency of rule revisions to adapt to local environmental and social changes

- **Collective-choice arrangements:**

  - Metric: Proportion of resource users participating in rule-making or decision-making meetings

  - Metric: Level of satisfaction with inclusiveness and fairness of decision-making processes

- **Monitoring:**

  - Metric: Percentage of resource users involved in monitoring activities

  - Metric: Frequency and coverage of monitoring patrols or checks

- **Graduated sanctions:**

  - Metric: Number and types of sanctions applied, categorized by severity

  - Metric: Perceived fairness and effectiveness of sanctions based on user surveys

- **Conflict-resolution mechanisms:**

  - Metric: Availability and accessibility of conflict-resolution forums or mechanisms

  - Metric: Average time taken to resolve conflicts, and proportion of conflicts resolved satisfactorily

- **Minimal recognition of rights to organize:**

  - Metric: Legal recognition status of the community's governance body by external authorities

  - Metric: Level of support or interference from external institutions, measured through stakeholder interviews

- **Nested enterprises:**

  - Metric: Number of governance layers involved in managing the resource

  - Metric: Frequency and quality of communication among different governance levels

  - Metric: Presence of coordination mechanisms or joint management agreements

These social metrics can be measured through mixed methods, including surveys, interviews, observations, and document analysis, to capture both quantitative and qualitative aspects. They provide a way to empirically evaluate how well a community adheres to Ostrom's principles and where improvements might be needed.

Urban Development Without Expansionist Reflexes

For the past 100 years, growth and expansion have shaped how we plan cities, manage finances, and perceive progress. Plateaus require a new mindset. The main task is to focus on what we already have. This focus is not just about technology. We must include values and the overall appearance and experience.

First of all, we must repurpose what is empty and underused. Here are some ideas to start the work:

- Former office towers can be converted into homes and live–work spaces, with ground floors that host clinics, elder daycare facilities, maker shops, and libraries.

- Former retail big-box stores can be repurposed into schools, production kitchens, training centers, and fabrication labs.

- Tight zoning that blocks modest additions and conversions in low-rise neighborhoods should be loosened in exchange for enhanced affordability and climate performance.

- Surface parking lots should shrink or vanish as on-street parking gives way to busways and protected bike paths.

- Streets should be treated as public rooms, with less asphalt and more trees, shade, benches, and fountains to combat heat and flooding.

At the same time, the city should acknowledge the need to reduce some assets gradually:

- Schools may have to merge as cohorts fall.

- Public authorities should plan dignified transitions: no locked, decaying buildings, but planned handoffs of facilities to new uses, with attention to memory and community.

What cities must not do is declare whole neighborhoods "abandoned territories" in the name of efficiency. People have histories, and land has meanings. If de-intensification is on the table, it must be voluntary, co-designed, and fairly compensated.

Critical services, such as water, health, connectivity, and safety, must remain available to those who stay, even as the city rebalances its population. Rewilding is appropriate in some places, but only with the free, prior, and informed consent of Indigenous communities, clear tenure, and governance arrangements that respect their leadership and local livelihoods.

Energy, Compute, and Virtual Residence

Many people in a plateau will spend a large share of their waking hours in digital environments because those environments enable creativity, collaboration, and community at low material intensity. That change will have ecological and reliability consequences:

- Cooling solutions should be designed to prevent harm to surrounding aquatic ecosystems.

- The carbon footprint embedded in hardware components must be transparently reported and actively controlled.

- Thermal energy by-products should be captured and repurposed whenever possible.

- The regulatory framework outlined in Part I must mandate annual drills testing, black-start capabilities, and system islanding, encompassing not only data centers but also the critical civic infrastructure they underpin.

- The required graceful degradation modes will be crucial in a world increasingly reliant on digital life.

Because digital continuation consumes energy, a small continuation compute tithe is warranted. A fraction of the gross energy and compute spend by continuation service providers should flow to the Perpetual Conservation Trusts on terms set out in law, with transparent disbursement and rigorous audits. This is a way to recognize that virtual life must pay rent to the biosphere that supports the power and materials that make it run.

Geography and Pattern Without the Sprawl Reflex

Societies that have reached a plateau in population growth will still exhibit high levels of mobility. Dense, high-amenity metropolitan areas will continue to thrive when they offer proximity, culture, and shared services.

Polycentric regions of medium-sized cities, linked by reliable rail and fiber networks, will also perform well, especially if they maintain abundant housing by building within existing footprints and reusing existing structures. While virtual residences will alleviate demand in certain areas, it will simultaneously strain electricity, cooling water, and transmission systems; therefore, location choices must consider local water and energy capacities to avoid causing concentrated environmental burdens.

Beyond the core regions, habitats will gradually become fragmented. Restoring natural landscapes and creating wildlife corridors will significantly enhance species diversity and improve carbon capture. However, without clarity on tenure, guardianship funding, and service guarantees, "rewilding" can become dispossession and neglect. Conservation works best when the people with long memories lead, and when the revenue for stewardship is secure across centuries.

Cultural Evolution Without Generational Churn

The scariest risk in a plateau is not underbuilding or overbuilding to adapt to the cities. The problem is not bricks or concrete but the impact on the mind. When people do not die, their minds can enter a state of stasis because their duties never end, and the system can take a long time to develop norms to accompany these changes.

We will discuss the mechanics of constant regulation and update so the legal system remains fresh in future chapters. Nonetheless, we still need to set up some connections. We have explored some adaptations in the workplace, and they might apply here as well.

Leadership roles should rotate, with sabbatical breaks and term limits. Instead of random selection by lottery, citizens should gather in assemblies to review significant policies and social programs. Professional practice should be renewed through recertification and portfolio reviews, not as a punishment, but as a means to enhance and update professional performance.

The public media will play a key role in replacing algorithms in digital realms. The media should be in charge of showcasing the new dynamics and giving a voice to minorities to prevent induced bias. The way in which art is expressed will also change. Cultural institutions should curate a long memory and invite reinterpretation of existing creations rather than building mausoleums or cold exhibitions.

Even though it is a complex process, renewal also requires making time for play. Cities can offer programs that give older adults a few protected years to try new things, without stigma or penalty, which will help people avoid getting stuck in routines. Innovation events can enable communities to review and update key programs and traditions, with clear guidelines for proposing and funding. These actions will keep alive the sense of community and of moving forward that might seem to fade as life endures endlessly.

Failure Modes and How to Mitigate Them

Systems optimize what they measure. If you measure only unit costs in housing, you will obtain units that are cheap but leaky and substandard. If you measure only the expansion of lane miles, you will get road deaths and heat islands. If societies fail to measure black-start performance, they will get promises and preventable flaws. A plateau society must avoid simplified, unidimensional metrics.

I want to emphasize the importance of metric pluralism as a choice architecture that guides decision-makers away from narrowly focused cost metrics. Immortal societies need to rely on multicriteria dashboards, independent audits, and the ability to say "no" even when a project looks cost-effective. These are not luxuries. They provide the behavioral cues necessary for robust sufficiency over fragile efficiency, aligning with insights that support better policy defaults.

There will also need to be social failure modes. For instance, digital divides might harden. To counter them, we can build compute commons and make high-fidelity virtual spaces accessible, just as libraries made books accessible a long time ago.

Other typical failures are backlashes driven by nostalgia and testimonials of scarcity. To counter these, we must emphasize the concrete, everyday benefits of a maintenance-first policy, ensuring they are visible and widely shared. By highlighting tangible improvements, such as the greater number of shaded walks, quieter streets, and lower utility bills, we can create immediate, recognizable benefits that resonate with individuals.

Maintenance goes more unnoticed than innovation because no ribbon is cut. Still, you can combat this by establishing minimum maintenance shares and tying executive bonuses and political credit to state-of-good-repair indices. These strategies, although lacking in glamour, can prove effective in making benefits and well-being visible for people.

Furthermore, social norm campaigns and public recognition can play an essential role in reinforcing maintenance-oriented mindsets. For instance, community-driven projects could celebrate neighborhoods that excel in sustainability through awards or public events, thus nurturing a culture of appreciation for well-maintained spaces. Highlighting positive examples through local media and workshops can also motivate other areas to follow suit, creating a ripple effect of enhanced maintenance practices.

What to Watch

The evidence gathered from the metrics should be presented in the form of actuarial tables, building permits, budget documents, laboratory notes, and park signage.

On the demography side, the charts will include the following:

- cause-specific death rates from aging falling at older ages

- plateaus in all-cause mortality at high ages under the longevity stack

- persistent sub-replacement fertility and rising ages at first birth

- the number of births and the number of remaining deaths converging and staying within a narrow band over many years[3]

On the economy side, the metrics will depict the changes below:

- maintenance taking the majority share of infrastructure capital spending in mature places

- retrofit permits outnumbering expansion permits

- "state-of-good-repair" indices improving rather than declining

- higher take-back rates for targeted product categories and steady increases in the economy-wide circularity indicator [4]

In urban form, the signals will include:

- permanent bus lanes and protected bike corridors replacing general-purpose lanes

- the number of office-to-residential conversions rising then stabilizing

- relaxed parking minimums followed by parks and housing on former lots

- travel surveys showing more short trips by active modes and transit without loss of access to work and care[5]

In energy and computing, the relevant data will be as follows:

- water withdrawals and return temperatures

- clean power shares

- embodied carbon

- the results of reliability drills overseen by a regulator you can name

- the continuation compute tithe collected and disbursed to biodiversity projects, including names, maps, and measured outcomes[6]

In culture and governance, the data will show the following:

- rotation and sortition mechanisms

- professional recertification

- sabbaticals taken without stigma

- innovation conventions producing specific changes

- cultural institutions curating a long memory while making room for reinterpretation

Closing: A World for the People

A plateau does not mean history stops. It means we can no longer run on autopilot. When population growth slows and people live longer, the old habits of building out, expanding, and chasing growth become less effective. What matters instead is how carefully and wisely we manage what we have, how fairly we share, how we stay within limits for resources, how openly we report progress, and how willing we are to share power and keep skills fresh so cultures remain vibrant.

The planetary boundaries and stewardship dashboards from Chapter 4 turn into local budgets and audits. The verification, consent, and provenance from Part I keep digital continuation honest and accountable. The principles we outlined in Part III about identity governance, continuity trusts, and "one identity, one vote" help prevent civic life from being manipulated by individuals who can be in multiple locations simultaneously.

Suppose you lay these rails and hold to them. In that case, a population plateau will become an opportunity to retrofit better than you built the first time, to restore ecosystems at scale, to design a digital life that is clean and reliable, to shape work and learning as a lattice rather than a ladder, and to let cultures grow old without growing stale.

Certainly, this doesn't mean problems will end. Next, we'll address how to move after we face and implement strategies to deal with scarcity. The sooner we confront the problems, the more possibilities we have to succeed.

Chapter 8

Economy After Scarcity

Why "Post-Scarcity" Needs a Careful Definition

"Post-scarcity" is often described as a world where money becomes obsolete, jobs are no longer necessary, and anything you want is instantly available.[1] This is an appealing idea, but it does not align with the realities of physics or the institutional limitations we discussed earlier.

For example, consider a common misconception that in a post-scarcity world, infinite energy and space would be available to everyone, allowing for limitless growth and consumption. In reality, the laws of thermodynamics and finite planetary resources fundamentally restrict such possibilities, necessitating careful management of what remains scarce. We have already covered the ecological limits, and reliability becomes even more crucial as technology becomes more affordable. Our starting point was a definition acknowledging that scarcity still exists and identifying its remaining presence.

In this book, post-scarcity refers to the increasing ability to provide more basic needs at almost no additional cost, reliably and equitably, without exceeding ecological limits. However, some essential resources will remain limited and need careful management. Take, for example, a family living in a suburban neighborhood that has recently been equipped with solar panels, a shared composting system, and community-maintained green spaces. Clean energy, recycling, automation, and shared computing have lowered the cost of essentials like housing, food, water, and connectivity, making them nearly free to use. As a result, the family now spends less on utilities and can afford to spend additional time on community activities and hobbies. They benefit from a lifestyle that encourages sustainability and equitable resource distribution. The effort has shifted to building, maintaining, and managing the systems that make this possible.

That is, scarcity does not disappear, but it shifts to areas such as time, attention, energy, reliability, computing power, maintenance, material quality, space, care, and trust. As such, post-scarcity is not about removing limits but about shifting them: from price to reliable access, from individual buying power to public reliability, and from competition to shared standards with protections.

What Remains Scarce After Abundance (And Why)

Time and attention remain hard limits even for immortals. Herbert Simon's observation from half a century ago remains accurate now: Information consumes attention, and an abundance of information creates a poverty of attention. You can automate food preparation, heating, transport, and diagnostics, but you cannot automate meaningful attention paid by a mentor, a clinician, a parent, or a friend without loss.[2]

In a world of digital continuation and long lifespans, the number of simultaneously held commitments that feel real is finite. Elinor Ostrom's principles for managing common-pool resources, discussed in Chapter 7, can be insightful when connected with governance principles such as clear boundaries, congruence with local conditions, and collective-choice arrangements, especially in a future world where people live very long lives and are hyperconnected. By framing attention in this way, we can ensure coherence in mitigation. Here's how these ideas might intertwine:

- **Clear boundaries:** Ostrom emphasizes clear boundaries around resource systems and the individuals or groups with rights to use them.

  - In a hyperconnected world with extended lifespans, clear boundaries could extend beyond physical or geographical limits to include digital realms and social networks.

- ○ Governance systems would need to precisely define who holds rights and responsibilities over shared resources in both physical and virtual spaces, accommodating fluid identities over long lifetimes.

- **Congruence with local conditions:** Ostrom's principle argues that appropriation and provision rules should fit local social and environmental contexts.

  - ○ In a world of diverse, long-lived communities connected globally, governance must adapt to varying needs across time and space, factoring in both local traditions and global realities.

  - ○ This adds complexity but also richness, requiring multilayered rules that remain flexible to evolving conditions over extended lifespans and interconnectedness.

- **Collective-choice arrangements:** Ostrom stresses the importance of allowing most resource users to participate in decision-making.

  - ○ With hyperconnectivity, collective-choice processes could leverage digital platforms for inclusive, transparent governance, engaging people across generations.

  - ○ Extended lives mean governance systems must incorporate mechanisms for intergenerational dialogue and representation to maintain legitimacy and sustainability.

Overall, Ostrom's principles, when integrated with governance concepts in a future of long lives and hyperconnectivity, suggest a dynamic, inclusive framework. This framework would uphold clear rights with adaptive, locally tuned rules and participatory governance structures that are enabled by technology and mindful of intergenerational equity.

That is why this chapter treats attention as a governed commons, not a free-for-all market. Energy and the reliability of energy anchor everything else. You can proclaim that marginal costs are near zero, but if the power to run the system is fragile or dirty, your economics can be destabilized by brownouts. The top priorities on provision lists are clean electricity with storage, black-start capability, and islanding for hospitals, water systems, and critical data centers. As earlier chapters have argued, compute is "food" for many life-supporting services; without audited energy and drills, compute and networks are brittle, and "free" quickly turns into unreachable.

Compute capacity and connectivity are not magically infinite. Baseline cycles can be provisioned as a right, but low-latency, high-reliability capacity and frontier accelerators that stay competitive use a lot of energy. If millions of people simultaneously need to run high-fidelity models for clinical counseling, engineering, or creative synthesis, they will still need to stand in queues, and someone has to establish priorities. Other challenges are related to bandwidth and latency, which, if overloaded, will constrain how "real-time" shared experiences feel. Physical locality still matters for structure and for synchronization, and people will need to keep a balance.

New Problems, Tested Solutions

These new societies will inherit some problems from the past, such as materials and maintenance requirements. Circularity will reduce primary extraction and waste, but error, wear, and contamination will accumulate and must be remediated. Pumps will seize, filters clog, joints wear, insulation compact, and adhesives fail, no matter how careful people are or how innovative the technologies are.

If we skip maintenance in a world of "free" provision, we will unfold cascading failure. That is why we must reuse the "maintenance share" policy: Mature cities should devote the majority of infrastructure capital to state-of-good-repair, and "free" provision should freeze before that threshold is breached.

Place and biocapacity do not scale with desire. In particular contexts, land, watershed integrity, biodiversity corridors, and urban amenities often compete with one another. Digital life can dematerialize some of the demand for physical goods and movement, but it will not replace the benefits of green space, clean rivers, and walkable neighborhoods. Baseline policies that overlook the local context will perpetuate inequality.

I want to dedicate a few lines to people who devote themselves to embodied care and presence, who, even when augmented by devices and assistants, cannot be infinitely replicated without reducing them to shadows of themselves. A good nurse's shift, a caregiver's patient kindness, a teacher's attention, or a restorative hug are not infinitely scalable via software without losing their essence. The economic architecture must not only "value" care in rhetoric but also build rails that surface, credential, compensate, and protect it. Augmented versions will not, or should not, erase humanity.

Legitimacy and governance bandwidth are scarce. Societies can only process so many complex trade-offs in a given period. Attention to governance is part of the scarcity that persists after material abundance. Legitimacy takes time and transparency to build; once squandered, it is expensive to rebuild. That is why public dashboards, audits, and the ability to pause and roll back mechanisms are economic necessities, not niceties.

From Price to Provision: Baselines as Rights, and Money Where Scarcity Persists

If scarcity shifts to new areas, then economic policies must adapt accordingly. The main idea is to make basic needs available as public options with clear standards, and to let money continue to play a role where things are still scarce or premium. These public options are real systems with budgets for maintenance, energy, and materials, as well as regular reliability checks and open reporting.

Low cost is not enough if the service is unreliable, just as free access is not enough if it easily breaks down. Moreover, public investment can act as a catalyst for private innovation, transforming the baseline basket from a mere redistributive effort to an economically catalytic initiative. By tying maintenance budgets to industry's learning effects, there is potential for enhanced efficiency and innovation, benefiting both the public good and economic growth.

Consider a collection or grouping of different investment assets or projects bundled together:

- a safe shelter with adequate thermal control and clean air

- access to clean water and sanitation

- staple calories and micronutrients

- reliable mobility and transport for daily living

- primary and chronic health services with diagnostics and follow-up

- baseline computing and connectivity

- access to safe public spaces and green areas.

In many contexts, automation, modular manufacturing, and shared platforms can significantly reduce the marginal cost of a unit in that basket. But the marginal cost at the point of use is not the only cost that matters. Provisions must be anchored to reliability targets, ecological budgets, and equity dashboards. In practice, this means that city or regional authorities, cooperatives, and regulated providers must take on those baselines and open the front-door price point, while also disclosing the reliability, carbon, water, and maintenance pathways that make the front-door price credible.

Even with changes in consumption and how people supply their needs, money will not evaporate. It will persist wherever competing goods and services remain. What could that be? Anything!

- a prime balcony on a block lined with mature trees

- ultra-low-latency, high-availability compute for special projects

- handmade objects that carry a maker's touch and time

- instant, embodied presence on demand

- designs with extreme customization

Money will also persist as a convenient unit of account, bridging between different systems. But it will no longer be related to the domination of survival because basic needs will be covered in other ways. This doesn't necessarily mean that the economics of status and premium scarcity will be eliminated, because distribution may still be unequal. So, the ethics will be about keeping money from pulling survival back into its gravity well.

In this context of transition and new rules, institutions will not retire; they will repurpose. Risk pooling will move from private markets to mixed systems with shared ledgers. Regulatory responsibility will extend beyond bank balance sheets to encompass the reliability of essential services like electricity, water, computing, and healthcare. On the consumers' side, protection will face challenges in emerging areas such as algorithmic resource misallocation, identity theft within reputation systems, and addictive user engagement patterns. Public procurement will shift from merely purchasing off-the-shelf services to actively promoting the creation of shared resources. In every instance, the focus will be not on establishing new institutions but on integrating them effectively with existing frameworks.

A Map of Nonmonetary Value Regimes and Their Rails

If baseline provision is decoupled from price, how do people signal, coordinate, and reward value outside money? Let's consider four overlapping regimes:

- **Reputation and experience credentials:** Instead of using money, individuals earn and exchange value based on demonstrated skills, trustworthiness, and contributions. Reputation acts as a social currency, enabling access to resources, opportunities, or collaborations. This can incentivize cooperation and quality without the need for financial transactions, cultivating community trust and long-term relationships.

- **Mutual credit and time-based exchange:** Mutual credit systems allow participants to trade goods and services using credits that represent mutual trust, not cash. Time-based exchange assigns value based on time spent providing services, equally valuing everyone's labor regardless of traditional market prices. These systems facilitate direct reciprocity and encourage local economic resilience by keeping exchanges community-centered and interest-free.

- **Public-goods funding for commons production:** Resources are collectively funded to produce or maintain shared goods and services that benefit all, like clean air, knowledge, or infrastructure. Funding can come from pooled contributions, voluntary donations, or decentralized mechanisms. This regime moves away from monetary transactions for individual profit and toward collective stewardship and equitable access.

- **Attention and creativity markets:** Attention (such as views, likes, or engagement) and creative outputs become valuable currencies. People are compensated through recognition, influence, or access rather than money alone. These markets harness intrinsic motivation and social dynamics, rewarding innovation and cultural contributions in nonmonetary ways.

Together, these regimes can complement or substitute for money by embedding social value, trust, and cooperation at their core. They promote economic interactions grounded in human relationships, shared goals, and alternative incentives that are well-suited to hyperconnected, long-lived societies. None of them "replaces" money in an absolute sense, but together they diversify coordination and, when tied to rails, become less prone to abuse.

In Real Life

To transform these concepts from theoretical to practical, municipal leaders could initiate local pilot projects as small-scale tests of these regimes. For example, a city could experiment with a community-led time bank for neighborhood services, such as tutoring or local repair, accompanied by an easy-to-use digital platform to track and exchange time credits. In rural areas, such a time bank could be adapted to address agricultural tasks or artisan crafts, strengthening local sustainability and community ties. Internationally, a similar system could be beneficial in regions where monetary transactions are limited, facilitating new ways to exchange value that align with local customs. Additionally, in marginalized communities, these regimes could empower individuals by valuing cultural exchanges, traditional skills, and local expertise. Such initiatives would allow residents to engage directly with the mutual credit system, offering a tangible pilot for potential broader adoption.

Reputation and experience credentials are verifiable attestations of contributions in specific contexts, such as hours spent mentoring, patches merged into an open-source repository, biosafety laboratory competencies, caregiving delivered, community translation performed, and neighborhood retrofits supervised. The goal is not to create one score that follows a person for life, but to make important work clear, easy to share when needed, and limited to its specific area.

Last but not least, equity requires that invisible or stigmatized work be valued. Jobs such as moderation, translation, patient hand-holding, cleaning, and safety checks are contributions and need to be fairly compensated, not just with social esteem.

Mutual credit and time-based exchange systems enable people to create credit for one another directly. This is as old as neighborly help and as new as digitally mediated time banks, co-ops, and local exchange trading systems. In a post-scarcity context, they are most useful for services that are inherently local and competing, such as childcare, tutoring, home retrofits, errands, and community garden maintenance. The rails are about safety, reliability, and dispute resolution.

All offerings exceeding the minimal risk limit must be validated through official certifications or overseen by authorized experts. Systems must have lightweight arbitration with clear escalation paths for fraud or harm, and municipal or cooperative insurers must backstop errors with transparent liability paths.

Equity can be built into multipliers or public top-ups for historically undervalued labor, ensuring that care does not become the unrecognized backbone of the "free" economy again. Caregivers who have time bank credits must be able to exchange them for services, like subsidized childcare slots, or community resources, such as shared kitchens and laundry facilities.

Public-goods funding mechanisms, such as quadratic funding and retroactive rewards, enable people to steer resources toward open, nonexcludable goods, including open transit tools, firmware for low-cost sensors, shared datasets, park improvements, and community archives. Quadratic funding amplifies many small contributions rather than a few large ones, while retroactive rewards pay teams for demonstrated impact after the fact rather than for merely promising impact.

In both cases, identity rails and audits are essential. Contributors should authenticate using privacy-preserving credentials to mitigate Sybil attacks without compromising donor anonymity. Impact committees must rotate membership regularly and disclose any potential conflicts of interest. Public procurement staff can support these funding models by awarding extra points to proposals that contribute to shared resources and implement open standards. This results in dual benefits: enhanced public assets and more cost-effective, compatible bids.

Why Attention Still Matters, and How to Keep It Humane

The attention and creative industries allocate limited time resources to shape cultural and social experiences. They already exist; the question is how to govern them so they contribute to social progress and don't contaminate civic power. In a post-scarcity setting, concerts, salons, multiplayer games, studio residencies, and co-created artworks take full attention. In mature societies, creators can earn a reputation by hosting, moderating, mentoring, and making.

Even with baselines secure, culture remains the fabric of meaning. Attention, as Simon, Goldhaber, Davenport, and Beck have argued in their different ways, is both scarce and constitutive.[3] What we collectively look at shapes what exists in our shared world. The goal is to hold the line between culture and civic power and to make cultural markets less extractive and more generative. That is why this chapter draws attention to platform audits, exposure diversity, anti-gaming, mental health defaults, and energy and water disclosure. It is also why it separates culture from governance: Applause should remain applause, not ballots or licenses.

For this, we should keep civic parity intact. The "one identity, one civic vote" logic remains, and popularity must not be able to buy offices or multiply ballots. Platforms must build anti-gaming tools, such as rate limits on influence growth, anomaly detection, and deliberate "serendipity slots" where underexposed works can surface, and must also measure exposure diversity.

Mental health protections should be the default, with standardized attention budgets, quiet modes, and friction on compulsive patterns, with funded moderation and care for creators who face harassment and burnout. Finally, because attention platforms can be computationally intensive, they should disclose their energy and water footprints, contribute to the continuation compute tithe, and participate in reliability drills, just like any other life-supporting service.

Integrating With the Structural Policies

The economic layer cannot exist independently of the governance and stewardship layers. It must borrow and reinforce them. Identity governance is the first connection, because living with copies has practical implications for credentialing, reputation, and the funding system. Without these factors, Sybil attacks will swamp funding; with them, we can keep plural selves from multiplying their sway, and we can attribute promises and liabilities to the correct instantiation.

The next relevant connections are linked to energy and compute. If baselines are rights, then baseline compute and storage must be treated as rights. That discipline is how you keep a free front-door price from turning into a hidden environmental bill.

Related to this, we must consider stewardship dashboards. Every large provider, whether a public option or a platform, must appear on the city's or region's public dashboard with three key metrics: ecological budgets and compliance, reliability service-level agreements (SLAs) and drill performance, and access disparities by geography, race, and income. When gaps persist, budgets must move. When reliability fails, expansion must halt and corrective investments must be mandated. Ecological budgets might be tightened, so the revenue from pricing externalities should be allocated to the Perpetual Conservation Trusts rather than to general funds.

Finally, audits and preregistration are the fourth connection. Mechanisms that allocate attention, reputation, credit, and public funds must be preregistered and subject to audit, just like the other rails. Adverse events, such as fraud, harassment, inequitable outcomes, and ecological backsliding, must trigger postmortems and, when necessary, rollbacks. This is how a culture shifts from "move fast and break things" to "move deliberately and fix things."

A Phased RoadMap From Now to There

Because this book is a work plan, not a wish list, we need to focus on the path that lies ahead. The first phase is a mixed economy with targeted pilots. The second phase scales what works and hardens the rails. The third phase shifts baselines toward a near-zero marginal price while maintaining reliability and budgets. Below, we take a close look at each phase.

Phase 1: Mixed Economy and Pilot Experiences

In the first phase, cities and regions should expand public options where the case is strongest:

- Clinics can open near-zero-marginal-price lanes for standard diagnostics and chronic care, backed by clean power with storage and a compute commons window for imaging models.

- Transit agencies can accelerate bus rapid-transit and protected cycle networks to make short trips cheap, fast, and safe.

- Housing authorities can prioritize the retrofit of studios that bring older stock up to thermal, air quality, and accessibility standards.

- Community repair hubs and libraries of things can scale with procurement policies that prioritize repairability and provide access to parts and manuals.

At the same time, bounded reputation pilots can start in universities, hospitals, and open-source foundations:

- Universities can issue verifiable credentials to mentors and lab assistants for specific competencies under supervision.

- Hospitals can credential peer navigators and community translators.

- Open-source foundations can embed contribution attestations into developer profiles, with audits to prevent "metric gaming."

- Quadratic funding and retroactive rewards can be trialed for local public goods, such as open transit data, firmware for air-quality sensors, and urban tree canopy enhancements.

- Contributors can authenticate using privacy-preserving credentials to mitigate Sybil attacks.

- Independent auditors can report on potential manipulation.

- Mutual credit and time banks can grow around care and education, with clear safety boundaries and public top-ups for low-income caregivers.

- Municipal schedulers can integrate credits to allow individuals to purchase respite and access to labs and childcare.

Phase 2: Scaling the Pros and Addressing the Cons

In the second phase, standards for credentials and interoperability can spread:

- Verifiable credentials can be transferred across compatible institutions, subject to contextual limitations and consent.

- Public procurement can give weighted credit to bids that contribute to open commons and that commit to reproducible builds and open data.

- Cities can dedicate a fixed share of their capital budgets to retroactive public-goods rewards, evaluated by rotating, conflict-screened committees.

- Compute commons can be built out with affordability indexes and queue policies, with expansions gated by clean power contracts and water disclosures.

- Law and policy can codify separations between attention markets and civic power, setting caps on cross-domain influence transfer for reputation systems so that star power in one domain does not dominate others.

Phase 3: From Baselines to Expansion

In the third phase, the baseline basket can approach near-zero marginal price at the front door, subject to ecological ceilings and reliability SLAs. This is when the claims in "post-scarcity" slogans come closer to being true, but only because the rails hold:

- Money persists in domains where rivalry remains, where embodied presence and tacit skill command a premium.

- The maintenance share of capital budgets in mature regions must remain above the threshold, not because maintenance is glamorous but because neglect makes "free" fragile.

Counting What Counts

When transitioning from price to provision, it is essential to consider various factors. A dashboard geared toward provision rather than retail prices must, at a minimum, show the share of households accessing the essential basket at near-zero marginal cost and within reliability SLAs. That means you must calculate, in a comparable way, what proportion of residents receive safe shelter within certain parameters:

- thermal comfort targets

- clean water within quality targets

- staple calories and micronutrients without food insecurity

- basic mobility for daily patterns

- primary and chronic health care, with follow-up targets

- baseline compute and connectivity

All of these parameters must lack financial barriers and have published reliable performance.

A second class of metrics covers reliability and resilience. Black-start drills and islanding performance for hospitals, water systems, and data centers should be depicted in simple language: time to restore function, percentage of loads covered, number of drills per year, and corrective actions taken. Additionally, these dashboards should include the mean time to recovery from outages and the rate of adverse events during outages. Failure should be treated as a signal to improve, not as a reason to hide in shame.

A third class covers ecological budgets, and the essential basket's provision must be reconciled against these budgets. When budgets tighten, expansion must pause, with corrective investments taking priority.

A fourth class covers equity. Access and outcomes for the essential basket, for attention markets, and for reputation and funding systems should be disaggregated by geography, income, and race/ethnicity, with automatic budget triggers when gaps persist. These metrics can help identify deficits and represent expansions in people's capabilities and freedoms.

Highlighting the above shifts the focus to the added freedoms people gain, such as reliable heat, lush tree canopies, and accessible compute labs. It should not be possible for a city to claim "post-scarcity" while leaving a third of its neighborhoods with unreliable heat, no tree canopy, and no compute lab. Relevant authorities must conduct thorough assessments to identify areas of need and allocate additional resources. Community oversight committees will need to be involved to ensure that corrective actions are effectively implemented.

Finally, a fifth set of metrics covers culture and safety. Exposure diversity indices on major attention platforms, harassment and harm incident rates, and response times, as well as platform energy and water auditing reports, should be made public. Creativity is not just the output; it sets the conditions under which the output is made and shared.

Failure Modes and How We Mitigate Them

Every mechanism we have sketched has a failure logic in place. The economic architecture must treat those logics as first-order design inputs.

Sybil Attacks and Bot Farms

If identity is cheap to mint, then quadratic funding will be easy to scam, reputation systems will be easy to inflate, and automated cliques will be able to capture attention platforms. The mitigation is a combination of verifiable credentials anchored to branch registries and rate and velocity limits on influence growth. It also includes anomaly detection that looks for coordinated manipulation. Additionally, incident bounties can be used to pay independent researchers to find and report exploits. There must also be a willingness to roll back when manipulation is found and to claw back ill-gotten rewards.

Celebrity Lock-In and Inequality

Popularity begets visibility, and visibility begets popularity. In an ungoverned attention market, a few people and institutions can capture an outsized proportion of mindshare and resources. Mitigations include the following:

- progressive "attention taxes" to fund underrepresented creators and communities

- deliberate "serendipity slots" where underexposed works are surfaced by lot

- diversity-aware recommendations that measure not only engagement but also breadth of exposure

- domain caps that limit how much influence in one domain can be carried into others

None of these eliminates "winner-takes-most" dynamics, but they at least reduce the slope and carve out breathing room.

Mental Health Harms

The same loops that make attention platforms sticky can also make them harmful, and the same dynamics that make creation gratifying can turn it into a punishing experience. Attention budgets, quiet modes, and friction for binge patterns should be on by default.

Moreover, platforms should fund moderation and mental health support. For instance, public options should feature "quiet rooms," both physical and digital, where presence is curated and paced slowly. The psychology of immortality, which we'll explore in Part V, has a place in economic design because when time stretches, patterns either deepen grooves or widen options.

Erosion of Privacy

An economic architecture that rewards contribution and engagement is hungry for data. Guardrails here include differential privacy, on-device aggregation, data minimization, retention limits, the "right to be quiet," and purpose-bounded use, enforced by contract and law. To preserve privacy, systems need to prohibit the sharing of attention and reputation data containing biometric identifiers without explicit, revocable consent. The system can't afford to lose citizens' trust.

Civic Capture

Great creators and charismatic voices should have large audiences, but they should not be able to purchase civic power with their following. The mitigation is the "one identity, one civic vote" rule extended to other realms. It should be enforced through the same branch registries and with rotation and sortition for oversight boards. A culture that confuses applause with authority loses its bearings.

Ecological Rebound

When the front-door price of goods and services approaches zero, demand can rise fast. Without ecological ceilings and reliability thresholds, efficiency gains can lead to higher total consumption in the classic rebound effect. The mitigation is to tie expansions of "free" provision to explicit budgets and to run pricing or quotas on the margins that have the most significant ecological leverage. The continuation compute tithe is a related mitigation. It institutionalizes reciprocity between high-energy digital life and biodiversity repair.

How Labor Changes When Survival Is Decoupled From Jobs

The most dangerous sentence at this point would be "There are no jobs." However, this would be false in terms of both logic and experience. What changes is the link between survival and employment.

When the essential basket is guaranteed and reliable, the existential cliff at job loss recedes. Paid work will still exist where the following criteria are met:

- reliability and safety are critical

- embodied presence and tacit skills are valued

- premium scarcity persists

- people must steward systems and commons with responsibility and accountability

Beyond traditional employment, new forms of meaningful contribution will also emerge, offering purpose and dignity for human beings. These roles will include community mentors, who guide the next generation; sustainability advocates, who protect ecological balance; and creative facilitators, who nurture cultural expression. As society evolves, the variety of dignified contributions will grow, reinforcing that the decoupling of survival from jobs still places value on meaningful work.

For those transitioning from traditional roles, support systems such as re-training programs, educational opportunities, and job transition assistance will become vital. These resources can help individuals adapt to new functions, ensuring that their skills remain relevant and valuable in this evolving landscape.

At the same time, mission-driven production will expand. People will spend more time on open science, neighborhood retrofits, education, translation, care, and culture. Those contributions will be coordinated through reputation and experience credentials, mutual credit, and public-goods funding, rather than solely by wages.

The boundary between "work" and "life" will change shape but not disappear; nonetheless, rituals and institutions must adapt. For instance, in safety-critical fields, recertification must take place on five- to ten-year cycles, while in all fields, sabbaticals for research, care, and public service should be expected, with easy re-entry paths built so renewal does not punish. Above all, care must be counted as first-order work.

Vignettes From the Plateau

In a midsize city, life looks like this:

- The public option clinic network operates near-zero-marginal-price diagnostic lanes and chronic care check-ins.

- The building's roof houses batteries that are connected to a district microgrid.

- The clinic conducts quarterly black-start and islanding drills in conjunction with the water utility and a nearby data center.

- A bright, accessible counter helps patients generate and manage verifiable data consents, allowing their records to travel with them and be revoked as needed.

- Nurses earn credentials for mentoring.

- A cohort of community translators earns experience credits that open access to a health communications studio.

- A retired clinician builds their reputation by supervising a resident-led open protocol for asthma management, which is published under a permissive license.

- Via the public dashboard, the clinic's reliability metrics and ecological footprint are visible to anyone.

- When a drill reveals a weakness in a switchgear cluster, the maintenance budget is allocated before a public announcement is made.

On the other side of town, a care guild anchors a time bank for elder and respite care. Members maintain a ledger of hours, but it serves a purpose that extends beyond mere accounting: It also records competencies and recognizes safe transfers, dementia-friendly communication, and end-of-life companionship. It also binds those competencies to supervision, so newer caregivers are supported and patients are protected.

The city tops up the time bank for low-income caregivers, so relief does not depend on family wealth. A rotating panel of guild members addresses disputes initially, then refers them to a municipal ombuds office if necessary. Credits can be redeemed for public childcare blocks, shared kitchen slots, and quiet hours in a compute lab. The guild publishes utilization and relief metrics; when a neighborhood shows strain, recruitment and training flow there first.

Every quarter, the metropolis runs a public-goods funding round for open tools and shared amenities.

Privacy-preserving credentials enable anyone to contribute a small amount to projects they believe will benefit them:

- an open-source firmware update for e-bike safety

- a neighborhood park shade project

- a set of community water quality sensors with a shared dashboard

- a memory archive for a historically redlined district

A quadratic funding mechanism amplifies small contributions, an independent audit firm publishes a report on manipulation, and a rotating impact committee allocates retroactive rewards six months later to projects that fulfilled their promises. The city's procurement team gives weighted credit to bids from teams that have delivered in previous rounds and published their work. The compute provider for the round discloses its energy and water footprints and posts its black-start drill report; a fraction of its revenue is allocated to the biodiversity trust.

Every aspect of social life is seamlessly integrated and has reached a more mature level. Life changes, and human beings find innovative ways to pursue goals and grow.

Tying It Back to Ecology and Maintenance

The more "free" the front door of provision becomes, the more discipline the back of house requires. No baseline is free of budgets. Every expansion of near-zero-marginal-price services must declare the energy, water, materials, and maintenance budgets. Reciprocity should be codified: The continuation compute tithe is a way to link digital abundance and biodiversity repair durably, while transparency ensures it remains honest.

Closing: A Build Plan, Not a Wish

Economics beyond scarcity does not mean eliminating money or work. Instead, it means shifting where scarcity exists and changing how institutions operate. The focus should be on solving the most challenging problems, such as reliability, maintenance, ecological limits, fairness, and trust, by building systems that can effectively address them.

The practices discussed throughout this book, such as audits, identity management, shared computing resources, and prioritizing maintenance, are what make free access possible and support creativity. If you are working on these ideas, start with these foundations:

- Measure what matters.

- Support shared resources.

- Keep a clear line between culture and civic power.

The rest comes from careful work and ongoing improvement. Now, think about the first step you could take to turn this hopeful vision into a real, livable future. That is what will make our collective efforts impactful and participatory.

In the next chapter, we will focus on governance from the political perspective. When we talk about governance, we refer not only to political structures and policy. Politics implies discussing social conflicts and the shared search for the common good.

Chapter 9

Governance for Immortals

When Time Replaces Design

In Lena's city, the transit chair had held the position for so long that people started using his name to refer to the entire bus system. He was known for being capable and hardworking, and he never seemed to leave his post. The buses became more punctual each year, but the routes stayed the same. Every election cycle, new routes for growing neighborhoods and biotech campuses were showcased in glossy plans, but funding continued to go to older projects with established supporters.

Reporters joked that as long as the chair stayed, so would his plan. The joke works because it rings true: When leaders never leave, their ideas stick around too. If a democracy relies on people passing away to bring about change, it confuses natural life cycles with intentional design.

The Problem We Are Actually Solving

Without intentional succession, future leaders never emerge. Leader turnovers in democracies are crucial because they ensure accountability and prevent the concentration of power, ensuring a healthy political system. Regular transitions enable fresh perspectives and policies, reflecting the evolving will of the people and strengthening democratic legitimacy.

Modern democracies often depend on turnover that happens because people retire, lose elections, or pass away, rather than because of built-in rules. Generations come and go, parties rebrand to attract new voters, and laws are updated after significant events. Time itself is what keeps things moving. However, when people live much longer, this natural turnover becomes less pronounced.

This leads to three main risks.

- First, leaders may stay in power simply because they can, building up name recognition and networks over the course of many years. Historical examples such as Fidel Castro in Cuba and Robert Mugabe in Zimbabwe illustrate how extended tenures can entrench power and stifle new ideas.

- Second, if the population doesn't change much, the issues that matter stay the same, and political platforms stop evolving. In countries like Japan, where population change is minimal, policy stagnation is often a topic of discussion.

- Third, influence can accumulate as wealth and experience grow without new people entering the scene.

There's also a newer and significant risk: the power of large platforms. In today's digital age, immense power is held by a small number of large platforms that control the flow of information. These companies act as gatekeepers, deciding what content people see and when they see it. This concentration of control can distort public perception by shaping which issues gain attention and how they are framed. As a result, certain problems may be exaggerated, downplayed, or ignored altogether, influencing public opinion and policy responses in ways that may not align with actual priorities or facts.

This power imbalance challenges the ideal of an open and democratic information environment, where diverse perspectives contribute to understanding and solving societal issues. It can also lead to echo chambers, misinformation, and polarization, making it harder to find common ground or address problems effectively.

The dominance of these platforms raises important questions about transparency, accountability, and the need for regulation to ensure that information distribution supports informed decision-making and healthy public discourse rather than serving narrow commercial or ideological interests. Addressing this risk is key to maintaining a fair and functional democratic society. Today, a few companies control what information people see and when, which can distort how problems are noticed and solved. This happens quietly, without a single person being to blame.[1]

One solution is to set stricter limits. These might include term limits, mandatory election cycles, and legal restrictions on consecutive or lifetime presidencies and other public office positions. These measures promote political renewal, reduce the risk of authoritarianism, and help maintain democratic fairness and balance.

All of this helps to address the underlying issues, but it is not enough. Leaders staying in power is a sign that the system was never designed for regular renewal. Instead, we need a setup where renewal, fairness, and transparency are built in from the start.

The tools for this, like identity management, reliable infrastructure, public dashboards, and controls on attention, have already been discussed. Now, the task is to put them together to support democratic politics, set public goals, and practice making changes so that updating and fixing things becomes part of the regular gear.

A Recap of Basic Elements

Let's first refresh the concepts and rails we have already discussed that serve as the bedrock for a brand-new political system. The first thing we must consider is that identity governance comes first. In a world with copies and branches, the registry would be the backbone of civic administration. Every person would hold a registry entry, and every branch would carry a branch ID. One identity would get one civic vote, full stop. If a person has branches, the right to cast that ballot should rotate internally by lot or schedule inside that identity, with the rotation and a record of representation bound to the branch ID.

Office-holding and control rights would also be linked to the branch ID and added at the origin identity level, so that a single life history could not multiply its influence through replication. Likewise, conflicts of interest and campaign finance disclosures would reference the branch ID so that provenance, recusal, liability, and penalties attach to the specific instantiation. No matter which branch, it must refer back to the original identity, who is responsible for any legal action.

The next key aspect is infrastructure reliability. If a selection system fails under grid stress, it cannot be considered a reliable selection system. A reliable governance platform includes the entire chain, from registration to ballot casting, tabulation, contestation, and public records. It must meet the exact reliability expectations of hospitals and continuation centers: clean power contracts, black-start and islanding drills, a documented cybersecurity posture, and published carbon and water footprints.

When a drill reveals a failure path, the system must have the resources and mechanisms to fix the problem, post a public after-action report, and allocate budget to strengthen the weak point. It is a work of engineering, because we can see social networks as pieces of engineering with interconnected parts that need to be attended to simultaneously.

Then, we must focus on stewardship dashboards. No political system remains stable without legitimacy, which is what makes people trust their leaders. Legitimacy is a function of both perceived fairness and measured fairness, so the publicity of public acts is the bedrock of these systems.

Governance must therefore publish disaggregated participation metrics, including turnout variance by district and income, time-to-hearing for petitions and complaints, time-to-remedy after adverse findings, access to translation and accessibility services, and office tenure distributions by level of office. Those dashboards should be bound by budget triggers, so that when gaps widen beyond the agreed-upon thresholds, preset mechanisms reallocate resources to address them. That's the same "count what counts; move budgets when gaps persist" logic you saw in the energy and ecology chapters.

Finally, is it possible to build a firewall between the attention economy and civic power? I believe it is not. Attention and popularity have cultural implications, but they cannot buy ballots or multiply representation. Limits on cross-domain interactions, recorded at the branch ID level, can limit the extent to which social influence converts into civic participation. Public media must emphasize exposure diversity and care less about engagement metrics. Similarly to what we discussed about the economy and the need for equality, here we see that amplification doesn't translate to enhanced participation. Moreover, amplification should not work against equality.

An Architectural Overview: From Default to Design

We are ready now to outline the architecture of a democracy that meets the needs of a mature society. It entails a set of mechanisms that work together:

- scheduled rotation with time-to-live limits on office-holding

- stratified sortition that makes citizens' assemblies normal and respected

- sunset clauses and reversible policy design that force reauthorization based on evidence and rights impacts

- minority rights safeguards that prevent majorities from eroding fundamentals just because they can

- independent redistricting and representation audits that keep maps from hardening around incumbents

- algorithmic governance rules that make code legible, contestable, and reversible

Each mechanism must incorporate administrative prerequisites, such as credentialing pipelines, stipends, training, and conflict registries, and each must carry signals of success and failure that the public can read. The point is not to adopt every mechanism at once but to make each piece of the system specific enough that a city, province, or nation can implement it, audit it, and either keep it or roll it back with its journals intact.

Below, we explore these mechanisms in detail. You will notice that they inherit many characteristics and procedures from the existing system but are adapted to a new individual and likewise a new public realm.

Rotation and Time-to-Live Limits

Term limits become insufficient on their own when lifespans no longer constrain political careers. A more helpful construct is a time-to-live limit for posts and roles. The charter could, for example, set a 12-year term limit for chief executives (served in two 6-year terms), an 18-year total service term limit for legislators, and a 6-year term limit for committee chairs and positions with agenda-setting authority. We need to consider longer periods because the perception of time will change, and so will social cycles.

Time-to-live limits can be paired with mandatory cooling-off periods so that stepping back is normal, not punitive. They should be coupled with the same paid sabbaticals and skill refresh curricula we saw for workplaces and education, ensuring that service is compatible with a life lived for centuries.

Several real-world examples demonstrate the successful implementation of this approach. For instance, Iceland's 1995 constitutional reform mandated committee rotation every 6 years, resulting in increased legislative diversity.[2] Additionally, New Zealand's adoption of sabbaticals for elected officials in 2001 helped prevent burnout and enhance policy innovation.[3]

The administrative prerequisites are ordinary and non-negotiable:

- credentialing pipelines that allow newcomers to prepare

- civic service stipends to prevent mid-career entrants from being penalized

- recertification pathways to maintain expertise without indefinite incumbency

- conflict-of-interest registries tied to branch ID

The signals you will want to watch are the median and 90th-percentile office tenure, the Gini coefficient of tenure across seats, the percentage of posts rotating on schedule, and the rate at which former officeholders complete recertification within a year.

When analyzing public office and elections, the Gini coefficient can be a powerful tool to measure inequality or concentration in the tenure of officeholders across different seats. The Gini coefficient is traditionally used to assess income inequality, where 0 represents perfect equality and 1 represents maximum inequality. Applied to office tenure, a low Gini coefficient would indicate that tenure lengths are relatively evenly distributed among officeholders, meaning that many officials tend to serve comparable durations. Conversely, a high Gini coefficient would reveal that tenure is concentrated among a few long-serving individuals, while others serve much shorter terms, suggesting potential entrenchment or a lack of rotation.

Alongside the Gini coefficient, examining the median and 90th-percentile office tenure provides insights into typical and extreme tenure lengths. The median tenure shows the middle point of all officeholders' durations, illustrating what is most common, while the 90th percentile highlights the upper range, revealing those who have remained in office significantly longer than most.

Furthermore, monitoring the percentage of posts that rotate on schedule is crucial to understanding whether regular turnover is a genuine practice or merely a theoretical rule. A high rotation percentage indicates that offices are indeed changing hands as expected, reinforcing democratic norms of accountability and renewal.

Lastly, the rate at which former officeholders complete recertification within a year indicates how often former officials are reintegrated into public roles, which can also signal patterns of office retention or cyclical leadership.

Together, these indicators (median and 90th-percentile tenure, the Gini co-efficient of tenure distribution, rotation rates, and recertification compliance) provide a comprehensive view of whether officeholder rotation is a genuine norm or just a fiction, which is crucial for assessing the health and fairness of public governance systems.

Building Legitimacy

Elections alone do not guarantee responsiveness, especially when turnout is unequal and parties can survive on the loyalty of stable voter blocs. Sortition, in this case, offers a complementary route to representative deliberation. When implemented with stratified sampling and transparent methods, it yields citizens' assemblies that reflect the community's composition and can weigh complex questions after structured learning.

However, to avoid exclusion, biases, and mistrust, the design must be carefully outlined for transparency. It must include the following criteria:

- The selection code should be open and reproducible.

- Invitations should be accessible.

- Acceptance should be compensated sufficiently to eliminate income as a barrier.

- Onboarding should include a civic curriculum and a manual with guidance about procedures, rights, and duties.

- Expert testimony should be balanced and subject to objection.

Assemblies can be bound to agenda-setting power, providing binding recommendations within predefined guardrails. Also, they can exert authority for specific domains, such as redistricting or sunset reviews, that can be updated over time according to society's needs.

The metrics that matter are these:

- The assembly's composition matches the registry on salient variables.

- Acceptance and completion rates are high.

- Participants show knowledge gains.

- Recommendations are implemented or, when rejected, receive transparent and reasoned responses.

Ensuring Fairness Without Demographic Churn

In an immortal polity, rules must prevent lasting entrenchment by the first coalition to capture a majority. Rotation across agenda-setting bodies should be structured so regions and communities that would otherwise be stuck in the opposition cycle can be empowered through procedural power. The system should be able to invoke minority safeguards easily enough to deter predation and be narrowly sufficient to stay legitimate.

Provision dashboards should trigger funding whenever governance fairness metrics drift, because legitimacy is tightly coupled to perceived and measured access. If petitioners in one part of the city wait three times as long for hearings or remedies, unfairness is not a bug but has been built in as a feature.

A note about weighted voting proposals is warranted here, as they are likely to recur. The siren song here is to compensate for entrenched majority power by giving marginalized groups more weight. Where you can vary weight responsibly is within rotating bodies that manage agendas and floor time. Even there, the weights must be authorized at a constitutional level, bounded by procedural contexts, and paired with minority rights safeguards and periodic reauthorization. If you let weight creep into ballots, you will create an arms race of justified exceptions.

Sunset Clauses and Reversible Policy Design

When demographic dynamics change, as we have seen, policies and programs decay. Some of them outlive their use, others cause problems over time, and some should grow but can't get political support. In a government without natural mortality, you need to plan for a clear way to end things.

A first set of solutions encompasses the following:

- Attach a time-to-live limit to new laws and programs. Schedule reauthorization windows and specify the evidence standards that must be met to maintain their current scale.

- Design decision gates and exit ramps into large projects so you can change course without treating change as failure.

- Keep policy journals for major decisions: Preregister the rationale, the alternatives considered, the expected signals of success or harm, and the stop criteria.

Those journals will become the institutional memory that a changing set of humans used to provide by leaving office and writing memoirs. They will also provide material for after-action reviews when you roll back.

If you want a domain where reversibility and retrievability have been taken seriously, look at nuclear waste governance. Good programs in this field specify retrievability periods, treat irreversibility as a cost, and institutionalize the idea that future generations should be able to revise choices as their knowledge, values, and technologies change.[4]

Minority Rights Safeguards and Judicial Fast Tracks

Democracies are founded on the principle of majority rule, where decisions are made based on the preferences of the majority of citizens. This ensures that the will of most people guides government policies and actions. However, majority rule is not carte blanche: It is balanced with protections for minority rights to maintain fairness and prevent oppression.

In long-lived societies, majorities are likely to remain the same for extended periods, which creates a series of risks. Eternal majorities would be indistinguishable from soft autocracy in practice. Protecting minority rights in an immortal polity, therefore, demands procedures that are both accessible and bounded.

In contemporary democratic systems, minority rights are preserved through constitutional protections that limit the power of the majority and guarantee fundamental freedoms such as speech, religion, and assembly. Independent judiciaries uphold these rights by reviewing laws and government actions to ensure they do not violate minority protections. Additionally, checks and balances among branches of government, and mechanisms like proportional representation in legislatures, help give minorities a voice and prevent majorities from wielding unchecked power.

These measures could be adapted to address the potential issues arising in societies where majorities might remain unmodified for centuries. One such mechanism is a narrowly scoped minority veto window for rights-reducing proposals. This could come paired with supermajority thresholds for rights changes, even when a proposal otherwise commands a bare majority. A second measure is a fast track to judicial review, with guaranteed standing and calendar priority for affected groups, ensuring that justice does not become a nullity through delay.

The only ways to keep these safeguards from devolving into trivial obstruction are to bind them to a formal rights catalog, require impact assessments for new proposals, and publish decisions promptly with reasons that a lay reader can follow.

Key indicators to watch:

- **Rate at which vetoes are invoked:** Measure how frequently minority groups or institutions with protective powers (e.g., courts, ombudsmen, or upper legislative chambers) exercise veto authority to block legislation or policies that may infringe on minority rights. A higher rate could indicate active protection, while too low a rate might suggest weak safeguards.

- **Time to adjudicate:** Track the duration taken by judicial or review bodies to resolve cases involving minority rights claims. Faster adjudication times can reflect efficient protection mechanisms and quicker relief from potential rights violations, whereas longer times might expose minorities to prolonged risks.

- **Rights outcomes before and after adjudication:** Assess changes in the status or scope of minority rights resulting from legal or institutional decisions. Compare rights protections or infringements before intervention and after rulings to determine the effectiveness of the system in upholding minority interests.

- **Share of proposals narrowed versus killed:** Measure the proportion of legislative or policy proposals that are modified or narrowed to accommodate minority rights instead of being entirely rejected. A higher share of narrowed proposals may signify a system that seeks compromise and inclusion rather than outright dismissal, supporting constructive minority engagement.

Together, these measurable indicators provide insight into how well contemporary democratic frameworks preserve minority rights against majority overreach.

Redistricting and Representation Audits

Maps are important. Geography influences democracy by shaping political boundaries, resource distribution, and population diversity, which affect how democratic systems develop and function. Geographical distribution directly impacts political participation by influencing access to voting, information, and political platforms. The geographic clustering of communities with shared interests or identities can affect how people organize politically and mobilize, shaping the overall dynamics of democratic participation.

When voting areas are drawn to protect current officials or to divide groups that would normally choose their own leaders, it hurts democracy. Redistricting, or redrawing these areas, should be done independently and must be checked to make sure it's fair. The process should use a range of methods, share all data and code clearly with notes and versions, and test for fairness in terms of shape, keeping communities together, and being fair to different political and racial groups.

If there are problems, the system should go back to a previous good map. People should have a chance to review it and suggest different plans. The tests and results should be explained in a way regular people can understand. Legal steps to challenge the maps should be easy to use. You should track things like how compact the districts are, how many communities are split, fairness scores, legal challenges, problems found and fixed, and how fast those problems are fixed.

Influence Limits and the Attention-to-Civic Firewall

Influencers in democracies in the age of social media have a significant political impact by shaping public opinion, mobilizing voters, and amplifying political messages quickly and widely. They can introduce new perspectives, engage otherwise disengaged audiences, and create grassroots movements that challenge traditional political institutions. However, their influence also raises concerns about misinformation, echo chambers, and the concentration of power in individuals rather than formal democratic processes.

Overall, influencers can both enhance democratic engagement and pose challenges to informed, balanced political discourse. If an influencer can convert followers into frictionless access and permission to dominate agendas, you have recreated aristocracy with better lighting.

Influence limits, therefore, require some rules:

- public lobby logs with branch ID provenance

- domain caps that limit the translation of attention into civic access

- cooling-off periods for officeholders leaving posts so they do not immediately enter paid influence

The firewall is equally essential:

- One identity gets one vote.

- Branch multiplication cannot multiply ballots.

- Popularity cannot change ballot weight.

- Media obligations and platform transparency reports should emphasize and measure diversity of exposure.

To effectively enforce these limits, it would be necessary to implement a combination of real-time monitoring and post-action audits. The use of technology, including algorithms and machine learning, could help track interactions and identify potential breaches promptly. These systems should integrate with public lobby logs to ensure transparency and accountability.

Additionally, a dedicated oversight committee could be established to review possible violations and impose penalties as necessary. This committee would monitor cooling-off periods to ensure that individuals do not exploit their former positions to gain undue influence. Public reports detailing enforcement actions would be published regularly to maintain public trust and demonstrate the commitment to upholding these standards.

Governance and AI Safety

Code is already policy in many domains. When you let models apportion resources, flag fraud, route emergency services, assist in adjudication, or propose maps, you reach higher speed and a larger scale, but you also introduce new failure paths. The honest way to use algorithmic systems in governance is to do the following:

- Build them to be legible and auditable.

- Test them under stress.

- Attach human accountability to irreversible actions.

- Publish incident investigations that do not spare egos.

Concretely, this means reproducible pipelines with frozen environments, dataset versioning and documentation, bias diagnostics, and ongoing monitoring under domain shifts. It also includes independent red-team audits before scale-up, human-in-the-loop for decisions that would otherwise strip rights or change representation, and rollback protocols with public diffs. Liability must be clear so that harms do not fall into the gaps between the agency and vendor.

Oversight and accountability models must be robust in this landscape. Institutional arrangements, such as independent commissions or ombudspersons, could be established to oversee the role of AI in governance. These bodies should be empowered to enforce transparency, handle grievances, and ensure that algorithmic decisions comply with legal and ethical standards. By implementing such oversight structures, we can bridge the gap between technology and governance, building trust and accountability in automated systems.

Long-Term Horizons: How Immortal Polities Decide Across Centuries

Humans and institutions often discount the future for reasons that are not always wise:

- We prefer near benefits to distant ones, even when the distant ones are larger.

- We underweight tail risks because they are abstract.

- We underprepare for regime shifts because our models are trained on yesterday's data.

Long-lived polities cannot simply claim to have outgrown those biases due to their lifespan. They must declare their discount policies explicitly, publish scenario weightings that prioritize low-probability, high-impact futures, and adopt practices from domains that already consider the long term. Again, nuclear waste governance is instructive here.

A city that tells itself, in public, "We will use a social discount rate of x% for climate adaptation, and here's why; we will require pessimistic floor portfolios for water and power; we will treat irreversibility as a cost to be budgeted explicitly; we will design for reversibility and retrievability" is a city that can course-correct without calling every course correction a failure. The literature on why citizens and politicians discount the future is not encouraging, but it is clarifying: If you know where you will be tempted to underweight, you can engineer counterweights.[5]

This scenario demands institutional time-to-live limits and scheduled review windows. When foundational rules are written as if they must last forever, we can see the importance of amendment as part of the ongoing process. In ordinary societies, amendments are often used only in the wake of a crisis. Such a schedule is ineffective in a society where citizens live for centuries and recall the lawmakers who drafted each rule firsthand.

A better approach is to attach long time-to-live limits of 50 or 100 years to constitutional provisions, with scheduled review processes that include stratified citizens' assemblies and expert panels, minority rights impact assessments, and clear ratification thresholds. This method would require supermajorities for rights changes and ordinary thresholds for process improvements.

To enhance transparency and keep citizens engaged, the government must publish the diffs and debates. These records become part of the civic archive, allowing the subsequent review to build on memories and past experience.

Administrative Rails: The Unglamorous Backbone

All the mechanisms above rest on administrative choices that are easy to overlook. Let's analyze some of the most important:

- Compensation and access determine who can serve. If civic service does not come with stipends and childcare, transport, and accessibility supports, the pool of participants will narrow to those who can afford it.

- Credentialing and training determine whether the rotation is real. If the only people who can step into roles are those who have already served, you have not achieved renewal.

- Conflict registries and recusal prompts determine whether integrity is practiced rather than preached. These registries should be live, public, and bound to branch ID so that conflicts cannot be laundered through branches.

- Scheduling must respect sabbaticals and rotations, and must also maintain replacement pools, so that the organization does not treat absence as a crisis.

- Performance review norms must focus on craft rather than charisma; otherwise, attention will seep back into civic power through performance metrics.

Worked Example: A City's Year of Reform

Consider a midsize metropolitan region in the 2090s that decides to stop relying on obituaries to renew its institutions. It amends its charter to establish time-to-live limits on executive and legislative roles, with mandatory cooling-off periods and paid sabbaticals. It pairs those changes with a modular civic curriculum, recertification pathways, and stipends so mid-career entrants can afford to serve.

It passes an ordinance that requires citizens' assemblies, chosen by stratified sortition, for major capital plans and charter revisions. The rules are clear and available to anyone:

- The selection code is published.

- The demographic match index for the first assemblies is now posted alongside the acceptance and completion rates.

- The city documents knowledge gains and commits to implementing or explaining recommendations within 90 days.

The government creates a separate redistricting board to make multiple map options, share data and code, and test maps for fairness and community needs. Additionally, it sets limits that cause a map to be rejected. If a problem is found, the city goes back to the last safe map, publishes a clear explanation anyone can understand, and tries again with new rules.

Fixing issues takes weeks, not years. This can be ensured through different measures:

- A minority rights group can stop a public safety rule if it might allow unfair stops.

- A special court reviews the rule within two weeks, and returns it with a smaller focus and clear notes.

- A fairness dashboard is available, showing voter turnout differences, petition hearing times, fix times, and how long officials have served, all broken down by area and income.

- The dashboard has alerts, so when problems exceed set limits, money is given to support clinics that help people join civic services, language help, and childcare.

Let's explore this through an example. The election system runs a test with the hospital and water company to practice restarting and working alone if needed. The incident report shows a problem during the switch and how it was fixed. Information about energy and water is included. The lobby registry is updated to track branch IDs, and limits are set to stop the platform from overreaching in filings and meetings.

Every three months, transparency reports show how spread out the platform's influence is. After a year, office workers change jobs more often, job time differences have leveled out, citizens' groups finish their work with good results, fewer legal fights over district boundaries happen, and the list of rights shows fewer emergencies and more updates. These changes don't make the news, but they do help keep the system working smoothly.

Failure Modes and Drills to Survive Them

Despite the meticulous planning, systems fail, but there is a significant difference in the consequences if those failures happen in predictable ways. Optimization collapse happens when a single "best practice" becomes an unquestioned dogma that fails under a new regime. The key is to maintain diverse portfolios and fund a variety of approaches. Moreover, it is necessary to hold periodic innovation conventions where norms and procedures can be examined to protect those who dissent from retaliation.

Credentialed capture happens when the rules say people must have certified qualifications but only insiders can actually get them. We can stop this by making qualifications easier to reach, paying for ways to get started, and valuing service without focusing only on rare experience.

Quiet inequity happens when data reports don't lead to changes when something is labeled "AI-assisted" but really hides how it works inside. We can fix this by making systems checkable, keeping humans responsible, and running tests that find problems before they cause failures.

Popularity seepage is a constant risk where popular voices take over. We can lower this risk by limiting influence from any one area, encouraging diverse input, and refusing to treat fame as equal to real participation.

Runbooks, Without the Bullets

A citizens' assembly feels like a leap until it becomes part of the institutionalized practices. The process would be as follows:

- It begins with sampling that a statistician and a civics teacher could explain to a neighborhood group.

- Invitations are sent out in waves, accompanied by clear and concise explanations of time commitments, compensation, and childcare arrangements.

- Acceptances are logged transparently.

- Onboarding is treated like a short course: the basics of the system, the rights at stake, the topics at hand, how to read a budget line, how to weigh expert testimony, and how to disagree without grandstanding.

- Deliberation is paced, with alternating blocks of learning and synthesis.

- Minority reports are encouraged rather than punished.

- There is a facilitator whose role is to oversee the process, not just the outcomes.

- Outputs are drafted in ordinary language, with evidence grades attached, and they include rights, equity, and ecological impacts.

- The assembly's recommendations are placed on the docket with a timer, and the receiving body must meet a deadline to implement or publish a reasoned refusal.

- Following each meeting, the group reviews feedback, identifies obstacles that hinder attendance, and applies these insights to improve future sessions.

Redistricting audits rely on consistent procedures to ensure integrity. These procedures include detailed provenance records and deliver the code with verified signatures, so external teams can replicate the builds. The ensemble model is described clearly so local journalists can accurately report on its findings.

The system also anticipates rigorous stress tests and publishes its results alongside concise interpretation guides. A designated public comment period allows map submissions, welcomes feedback, and transparently outlines criteria for raising objections.

If irregularities arise, such as a district's compactness score falling outside expected parameters without a justified community rationale, the established protocol activates smoothly. The emergency measures are as follows:

- Roll back the maps.

- Publish a follow-up report.

- Restart the process.

The objective is not to create a flawless map but to develop a transparent, accountable, and responsive system.

A constitutional sunset review would need a similar flow. For this, the government must post the bundles of expiring provisions months in advance, provide assemblies with stratified sampling, and appoint expert panels that reflect diversity and address conflict checks.

The public registries and reports must showcase evidence on performance, equity, and ecological alignment. Expert teams would propose text for renewal, modification, or termination, accompanied by detailed records explaining the reasons and any alternatives. They should also establish the approval criteria beforehand, with stricter requirements for altering rights. Once the revision is approved, all changes and discussions must be publicly archived to prevent redundant reviews.

Public Narratives Versus Built Plans

The public conversation about immortality and democracy will continue to swing between threats and promises of bright futures. Some will warn about surveillance, misinformation, and hubris. Others will celebrate boundless choice and competence. Despite being contradictory, such narratives help to spark interest and remind us of the real risks; however, they are not sufficient to build a city charter, a national constitution, or a platform procurement.

What converts a narrative into a plan is specificity:

- a registry that makes identity legible and non-duplicable

- a time-to-live limit that forces departure without scandal

- a citizens' assembly that any willing resident could serve on and recognize as legitimate

- a rights safeguard that anyone can invoke

- an audit that anyone can run to see if a redistricting map is fair

- a drill that will disclose a failure when it happens

The virtue of an immortal polity is not that it is unbreakable, but that it can repair itself without waiting for time to do the work.

Concluding Thoughts: Renewal on Purpose

In the past, natural limits on life helped shape how systems changed. As those limits fade, we must design ways for people to transition into and out of leadership roles. Setting time limits for holding office and offering sabbaticals makes stepping down a normal process, while random selection and good training help new people step in. When a society builds these systems, it doesn't have to wait for people to leave to make progress. Instead, it creates a government that learns, adapts, and openly demonstrates its work.

To ensure these proposals are practical and adaptable, further empirical research and pilot programs should be considered to test their implementation. Comparative studies across different governance models could provide insights into the strengths and limitations of these approaches. Such research would help refine these ideas, making them more actionable and aligning them with real-world complexities. By setting these proposals in motion, we can explore what truly works to ensure democratic renewal and adaptation.

Now that we have explored the social realm that frames the experience of timeless life, we will delve into people's minds. A change in the timeline of human life awakens new ways of connecting with others, connecting with reality, and projecting into the future.

Part V: The Psychology of Eternity—Mental and Emotional Life When Tomorrow Never Ends

Chapter 10

Love in the Long Game

A Promise With No Sunset

On the morning of their fourth century together, Eli and Nora arrived in a coastal town neither had seen before. They came without linked memory bands, without the background hum of synchronized thought, and without any artifacts that could pull them into the gravity of their past. They had agreed, again, to meet as if for a first date. In the restaurant's soft light, with strangers' conversations washing like a tide against the walls, Nora asked a familiar question in a new voice. "If we lived only eighty years, would tonight feel different?" Eli smiled. "Of course. But we don't. So let's write a new first page, and see what it does to the next hundred."

Immortal love endures not through a single, unchanging promise but through continual renewal and growth. It thrives on regular updates, moments of connection, understanding, and shared experiences that offer new opportunities to fall in love again and again. This ongoing process keeps love alive, adaptable, and deeply meaningful over time. It survives by scheduled renewal.

Eli and Nora had learned, time and again, that immortal love lasts not by following one long story but by choosing to refresh it on purpose. This is the main idea of this chapter: When time stops forcing decisions or change, you must create a structure that keeps intimacy flexible, ethical, and alive. The power of all might be enough, but it doesn't have to be. Couples can engage in key renewal practices that support this. Let's find out what might help.

Diagnosis: The Paradox of Eternal Time

Mortality shapes how intimacy is perceived. It makes time feel short, pushes people to make choices, and sets a natural rhythm for beginnings and endings. Without it, time would stretch out, and drifting apart would become easy. Problems that once needed fixing could be put off again and again, until they never get fixed.

Where relationships today break from too much urgency, immortal ones are more likely to fall apart from neglect. If a couple knew they would live forever, they might face several unique challenges in their romantic relationship:

- **Eternal commitment pressure:** The idea of forever could create intense pressure to maintain perfect harmony indefinitely, leading to anxiety or unrealistic expectations.

- **Risk of stagnation:** Without the natural life cycle of relationships, the couple might struggle with boredom or a sense of monotony, making it harder to keep the romance fresh and engaging.

- **Changing individual growth:** Over infinite time, personal growth and changing interests could lead to diverging paths, testing the relationship's adaptability and resilience.

- **Conflict resolution:** Small disagreements might become magnified over time, as endless interactions could lead to the accumulation of unresolved issues if not effectively addressed.

- **Emotional fatigue:** The permanence of the bond might cause emotional exhaustion as the couple navigates an unending journey with inevitable ups and downs.

Overall, living forever could amplify both the joys and tensions of romantic relationships, requiring exceptional communication, continuous renewal, and flexibility to sustain love over eternity.

The couples who last will not be the ones who never argue or never change, but the ones who plan for renewal and repair and use technology with explicit consent, tracking, and the option to reverse decisions.

Coherent Actions: Relationship Rails That Work in Practice

Most immortal couples who endure will treat their partnership as a studio for love and a lab for governance. They will write a living charter beyond their vows. This charter would be a compact operating agreement that establishes a re-consent cadence for intimacy technologies and a quick "pause/revoke" flow. These couples would also log presence and route shared assets through simple trusts with sunset clauses. They would schedule cycles of togetherness and independence, much like athletes schedule training and rest. Finally, they would specify how the charter itself changes, including who signs, how long edits need to cool before adoption, and how diffs are preserved, because memory drifts even when intentions are good.

A living charter would have several elements. First, it would name the parties not only by name but also by their branch ID, if there are copies or distributed instances, so that the obligations bind the correct instantiation.

Second, it would describe "mode accommodations" for cross-mode couples, such as biological, digital, and hybrid, so all versions could feel comfortable and make the others feel comfortable too. This could mean a pledge to provide quiet, embodied space for digital visitors, captioned and paced virtual experiences for embodied partners, and "translation rituals" where each explains their sensory experience of the same event.

Third, it would set the pace for consent. Annual renewals would become rituals, and the ability to pause or revoke consent for merges, memory-sharing, and any synchronized cognition tools would be respected as well as expected. These permissions should be saved as different versions with a history of where they came from. They should also be locked for security and shared, because this history helps check and undo actions later if needed. To illustrate, a consent renewal clause might look like this: "We agree to revisit our consent for memory-sharing every 12 months, allowing us to pause or revoke it with mutual understanding and without repercussion."

Fourth, it should describe presence-weighted obligations. If one partner carried more caregiving, more travel, and more late-night logistics for months, the next quarter should see a rebalance without drama.

Fifth, it must define assets and compute obligations. Continuity trusts would hold shared archives and long-lived projects, would appoint independent trustees, and set sunset dates so that obligations do not last indefinitely.

Finally, it should lay out cycles: the decades of shared work, the seasons of parallel play, and the sabbaticals, both together and solo, that function like breath for the relationship.

Everyone can write vows. Fewer couples write these rails, and those who do tend to last.

Guiding Policy: Build Rails and Renew Them on Schedule

Much like aspects of social life, politics, and the interaction between human groups and the environment, love and romantic relationships also need rails. These rails should align with the system-wide governance we have outlined so far. Thus, we can consider the following as the starting point:

- Relationship agreements should reference identity registries, not just names, so that promises can be mapped to specific instantiations.

- Consent should be viewed as a process rather than a one-time signature, especially when technology enhances intimacy.

- The who, what, when, and why of intimacy tech should accompany consent agreements so that it is clear to both parties and revocable.

- Presence should determine obligations; care ought to follow the one who shows up.

All of this should renew on a visible flow, because in a world where "always" is an available option, "again" must be chosen deliberately. Therapists could play a crucial role in helping couples implement these principles by guiding the creation of living charters and facilitating renewal conversations. In a society where people can live forever, effective communication and agreements for couples would be essential to sustain enduring relationships. Here are some tips and agreements therapists might suggest:

- **Regular relationship check-ins:** Schedule frequent, honest conversations to discuss feelings, evolving needs, and any concerns to prevent issues from accumulating over time.

- **Agreement on personal growth support:** Commit to encouraging and respecting each other's individual growth, hobbies, and changes, recognizing that people will evolve over long lifespans.

- **Flexible conflict-resolution protocols:** Develop agreed-upon methods for resolving disagreements constructively, including pauses if emotions run high, to avoid lingering resentment.

- **Renewal rituals:** Create shared rituals or new experiences periodically to rekindle romance and adapt to changing circumstances, maintaining the relationship dynamic.

- **Transparent future planning:** Maintain open dialogue about long-term goals and aspirations, adjusting plans as life circumstances evolve to ensure alignment.

- **Space and autonomy respect:** Agree on boundaries regarding personal space and independence to balance togetherness with individual freedom.

- **Embrace forgiveness and patience:** Recognize that mistakes will happen over extended time frames, and commit to forgiveness and patience as ongoing values.

Here are some example phrases and exercises couples could use to practice effective communication and agreements in a society where people can live forever:

- **Regular relationship check-ins:**

  - Phrase: "Let's set aside time each month to share how we're feeling and what we might need from each other."

  - Exercise: Schedule a monthly "relationship meeting," where each person takes turns expressing their thoughts and emotions without interruption.

- **Agreement on support for personal growth:**

  - Phrase: "I support your interests and growth, even if they change over time, and I hope you feel the same about mine."

  - Exercise: Each partner writes down personal goals or new interests quarterly, then discusses how the other can support them.

- **Flexible conflict-resolution protocols:**

  - Phrase: "If we ever feel overwhelmed during a disagreement, let's take a break and agree on a time to revisit the conversation calmly."

  - Exercise: Practice pausing mid-conflict and using a safe word or signal to pause and cool down before resuming the discussion.

- **Renewal rituals:**

  - Phrase: "Let's try something new together every few months to keep our bond fresh and exciting."

- ○ Exercise: Together, create a list of shared activities or experiences you'd like to explore, then pick one to do regularly.

- **Transparent future planning:**

  - ○ Phrase: "As our lives evolve, let's keep talking openly about our goals to ensure we stay aligned."

  - ○ Exercise: Annually review and update a shared "life vision" document covering relationship, career, travel, and personal aspirations.

- **Respect for space and autonomy:**

  - ○ Phrase: "I value our time together, but I also understand the importance of personal space. Let's check in if either of us feels crowded."

  - ○ Exercise: Agree on signals or cues that indicate the need for alone time, and practice respecting them without taking offense.

- **Embracing forgiveness and patience:**

  - ○ Phrase: "Mistakes are part of our journey. Let's prioritize forgiveness and patience as foundations of our relationship."

  - ○ Exercise: Reflect together on past conflicts and discuss how forgiveness has helped you grow, reinforcing a culture of compassion.

These approaches can help couples navigate the complexities of eternal relationships by cultivating trust, adaptability, and continuous emotional connection. They offer support in aligning relationship agreements with personal and shared goals, ensuring that the infrastructure of the relationship is both flexible and rooted in mutual consent and understanding.

Designing Consent for Intimacy Technologies

The technologies this book considers, including shared memories, merges that temporarily blend cognitive streams, and synchronized thought that accelerates co-creation, can enrich intimacy, but they can also wound trust. They require more than affection.

If you choose to adopt merges as a practice, treat them with the same seriousness you would bring to safety in a clinic. Merges can evoke a range of emotions: the thrill of deeper connection, the relief of shared burdens, or the vulnerability that comes when barriers fade.

Some things to consider start with purpose and scope:

- Ask why you are doing this now and what is out of bounds.

- Set a hard time limit for any session and a safe word that stops the process without debate.

- Isolate identity so that merges cannot alter stored values or long-term memory.

- Incorporate routine health assessments that automatically suspend activities during periods of excessive sleep deprivation, illness, or diminished cognitive focus, preventing poor decision-making.

- Protect third parties at the system level. Filters should prevent memories involving others from entering the session without their explicit consent.

After the session, debrief in the same way clinicians debrief procedures:

- Did you achieve what you hoped?

- Did anything feel off? What changes next time?

Store only what you must, such as cryptographically signed consent, session times, and participants, and agree on retention or deletion ahead of time; do not share the content of the session. Finally, schedule periodic sessions with a neutral counselor or ethicist. This doesn't mean you cannot be trusted, but devices that amplify intimacy deserve independent safety inspections.

The same care should apply to memory-sharing. Memory implicates more than the self, so it is advised that, before sharing, you obtain third-party consent or request redaction. Take the following into account:

- Opt to mask names and locations by default unless there is an explicit reason to disclose.

- Keep retention short by default, and require active, mutual renewal.

- Treat shared memories as "derived records" if they show up in therapy or adjudication. They can deepen empathy, but they cannot be treated as decisive evidence without corroboration, because merging and sharing changes recall and senses.

- Finally, make revocation real. Either partner should be able to revoke access later, and the system must propagate that revocation to any copies.

Measuring What Matters Privately

The book repeatedly insists that claims should be counted only when they change hard outcomes. In intimate life, the "hard outcomes" are connection, fairness, repair, and safety. You could build a simple private dashboard with a set of measures you review quarterly to help make the proper adjustments.

Each dashboard item should be linked to a specific, observable action, like "Repair within 48 hours" to ensure prompt resolution of conflicts, or "One novelty every 90 days" to maintain excitement and prevent habituation.

You would therefore be able to track whether you are honoring your flow of togetherness and independence. It is easy to declare an intention and difficult to keep it in the friction of real life. Tracking novelty is about ensuring that something genuinely new appears in your shared calendar frequently enough to keep things fresh and exciting.

By embedding the dashboard into your shared life rhythm with intentionality and flexibility, it would become a powerful tool to nurture a long-lasting, vibrant relationship. You could use a digital tool or an online template, but this information does not belong on a social feed or in an app store. It belongs in a private ritual, a quarterly conversation that begins with gratitude and ends with one concrete commitment each for the next quarter.

Sustaining Passion Across Centuries

You could take steps to avoid falling into a routine. One way would be to plan a regular "first-year reboot." For a set time, maybe three months every decade or so, you bring back some of the things that sparked that early passion: curiosity, new questions, time outdoors together, and simple, caring rituals at the start and end of each day. This isn't about pretending to be younger, but about bringing fresh energy on purpose.

Another idea is to take turns leading new experiences. One partner plans for a while, then the other takes over. Each suggests a few options, not to compete, but to make sure both have a say. Sabbaticals, whether together or alone, offer something unique. When you take one together, try something where you're both beginners, like learning about the ocean or music from another era. Solo sabbaticals work best when you share what you learned and bring back a story or a small change to your daily life.

Curating memories helps, too. This doesn't mean erasing the past; instead, you let go of routines that no longer serve a purpose and focus on memories that truly matter. For example, let go of that old argument you've replayed too many times, but keep the memory of your child learning to swim vivid. Taking on new challenges together, with some risk but clear limits, also keeps things fresh. The key is to choose risks that are exciting but safe, not ones that could harm your future.

Asynchronous Growth Without Breaking

Immortal partners would not grow at the same pace, and that is good. One of you might take a turn into a new field or mode and accelerate; the other might delve deeper into something that requires slow attention. This would only be a problem if one of the partners stopped bringing the other into the changes .

To deal with this, you could engage in regular growth audits. Once a year, sit for an hour and answer a small set of questions:

- What did I learn this year that changes what I need from us?

- Where do I want to run faster, and where do I want to run slower?

- What do I need to stop doing to make room for your growth?

Each partner should commit to one change and schedule a check-in in three months.

Echo sessions would use the way your family already shares memories: After exploring on your own, you share short stories or sensory notes, respect others' privacy, and ask for thoughts instead of agreement.

You would renegotiate boundaries when they needed renegotiation, maybe around exclusivity, cohabitation, or the scope of the hub. Then, you would write those changes into the charter, with cooling-off periods and sunset dates, so that nothing "updated" becomes permanent by inertia.

You could also schedule reconvergence windows: regular, non-negotiable periods every few decades during which you live together and reset. Think of your relationship like a collection of different parts. It includes daily habits, yearly traditions, long-term goals, and lasting memories. You review and adjust it every year because doing it once is never enough.

Commitments That Renew Rather Than Crumble

"Till death do us part" made sense in a world where death arrived without your permission. In a world where departure is elective, it would be better to write renewable commitments and to make renewal a practice rather than a slogan.

Couples could choose intervals that are as long as they wish: 25, 50, or 100 years. The relationship would continue not by default. Instead, it would demand active choice and expressed commitment.[1]

Cohabitation would become a module you could add or remove without dissolving love. The relationship would not be propped on a single living arrangement. Couples could negotiate alternative cohabitation by openly discussing and respecting each other's needs for both shared and personal space, creating flexible living arrangements that balance togetherness with independence. Setting clear agreements on time spent living alone would allow partners to recharge individually while maintaining a strong, supportive connection.

Parenting would change, too. In biological parenting, you would plan for presence-weighted obligations across long windows and specify tie-breaks for decisions without pretending that infinite time makes choices light. Where digital or hybrid children were present, you would align compute and bandwidth obligations with caregiving, and you would specify guardianship across branches so that continuity is not a gameable abstraction.

Ending Without Erasing What Was Good

In immortality, endings would be part of the architecture, but the choice to end would carry a different weight when there is no natural closure. That is why it is worth specifying exit ethics. There would be a duty of care; after two centuries together, a clean break would also be a fantasy.

One practical approach could be a closure sprint: a six-week period in which you update the charter for exit and route shared assets into a continuity trust with an independent trustee and a sunset clause. It would also be advised to schedule two reconciliation reviews to adjust the terms as reality reveals what the plan missed.

Therapists could play a pivotal role in facilitating these closure sprints or rituals. By guiding clients through the process, they could help ensure that decisions are made with dignity and clarity. Their involvement would support an ethical and compassionate practice.

To illustrate how a closure ritual might unfold, imagine a couple embarking on this journey. They might start by jointly reflecting on their shared history, acknowledging both joyful moments and challenges. Together, they would draft a goodbye letter, expressing gratitude and articulating hopes for their respective futures. Next, they could plan a ceremony involving close friends, where both partners share a symbolic act, like planting a tree to represent growth and new beginnings.

The archives a couple has shared would deserve protection. Two people could keep the shared records in private unless both agreed to share them. Separate records could be organized so others can understand the past without needing every private detail.

Shared memories should follow the agreed-upon rules for sharing and keep protections in place. If there are children or shared workspaces, responsibilities would continue, and you should clearly know your role instead of acting like the partnership's end cancels those duties. For legal decisions, shared memories should be treated as secondhand records. Courts should ask for proof and use careful checks, like comparing behavior, confirming personal memories, checking personality traits, getting validation from others, and tracing the source, to make sure someone is not faking the memories. Lastly, a small ceremony, like a "living wake" for the relationship, would help people feel both sadness and gratitude. It would also show how to end things without bad feelings.

Technology as Amplifier, Not Arbiter

Shared memory, synchronized cognition, and distributed identity are powerful amplifiers of collective intelligence. They are not arbiters of authenticity, and they can play relevant social roles:

- Use shared memory as a practice for empathy and storytelling.

- Avoid factory-scale sharing that turns your private life into a product.

- Keep retention short by default, obtain third-party consent first, and never share children's memories without their explicit, age-appropriate consent and the right to revoke.[2]

- Use synchronization for work and art that benefit from flow: composing, co-designing, co-surgery, if that is your life.

- Avoid making "always on" sync the default at home. Choose a small, stable quorum of instantiations for relational life.

- Do not ask your partner to relate to an unbounded swarm.

Any claim of continuity made by tech should be held to the same verification and provenance standards this book applied to digital continuation. If a device that creates closeness says it can keep or rebuild the person you promised to be with, it must pass set tests checked by others. If not, just enjoy the simulation for what it is (nice and helpful), but don't act as if it replaces a real promise.

Working thoughts together smoothly could hide the risk of losing personal freedom over time. Consider this scenario: Two partners engage in prolonged synchronization to enhance their shared creativity. Over time, subtle shifts occur in their personal values and preferences, influenced inadvertently by the other's thoughts. What began as a voluntary merging threatens each partner's distinct sense of self. They prevent this drift and ensure autonomy is cherished, reminding us that safeguards are nonoptional. The potential loss of agency illustrates why synchronization should be approached with caution.

A ToolKit, Told as a Story

Some couples treat their quarterly review like the sweetest kind of meeting. They begin with gratitude, each naming something specific the other has done that matters. They review their presence ledger and compare the numbers to their sense of fairness, then reassign a few responsibilities with less drama than it took to install a dishwasher.

They look at the calendar and choose one small novelty for the next three months and one larger project for the next year, so that their future is not just a procession of tasks. They ask, gently, whether any consent needs review. They pick one conflict pattern they have repeated, rehearse a different response that is kinder and more specific, and write the cue on a sticky note. They conclude with a sentence each about what they will choose next quarter, because saying it out loud matters.

Every decade or two, they run a "first-year reboot." They reset a few rituals, and they adopt a "mode day" where they each spend time fully in the other's primary mode, whether that's fully embodied, fully digital, or hybrid, and then write a paragraph about how it felt and what they want to change in their default hospitality.

They exchange letters about who they have become over the last decade and what they're asking for next. They look at the list of anchors from the last era and decide on three new anchors they want to add for the next chapter. They understand that love keeps its weight not by refusing to change but by choosing changes together.

Three Brief Cases

Elion and Karo have been together for 6 centuries. They learned early that their taste for novelty, which made the first 100 years feel like spring, would eventually work against them if they chased it without rhythm. They adopted a 50-year sabbatical pattern. Between reunions, they blur the day-to-day routines that can turn into contempt by repetition, but they never blur the anchors that make them a "we." On reunion, they read those anchors and add three new ones to plot the next era. Over time, they find that wonder lives on the edge between the known and the untried, and that edges can be found, or made, in any century.

Samira and Jo love each other, but they do not love at the same tempo. Samira has traded identities a few times over the past two centuries: aesthetic, then engineer, and then ethnographer of virtual microcultures. Jo collects languages and lives through long analog days that stretch like quiet rivers. Their growth audits surface an old pain: Samira's acceleration reads like judgment to Jo, while Jo's savoring reads like indifference to Samira. They write a scheduling bargain. On Sunday mornings, Jo pretends they are teaching 10-year-olds and reads a poem slowly until Samira can taste it. On Wednesday nights, Samira takes Jo to a new micro-scene and lets them absorb speed in a bounded dose. Twice a year, they ask whether to intensify or simplify their approach. The ledger looks balanced. More importantly, so do their faces.

Laine and Petra built a clinic that teaches couples how to use intimacy technologies safely. After two centuries, they reach a renewal window and see that the next honest move is to shift into friendship. They run a closure sprint: They update the charter, pour shared assets into a continuity trust with a 20-year sunset clause, draft a note for family and colleagues that explains without blaming, and schedule the two reconciliation reviews that reform has taught them to expect. The clinic carries on. Love becomes a cherished archive and a new practice of care. Their students learn that ending well is possible.

Why These Rails Matter

It's easy to get caught up in big ideas or poetic language when talking about immortality. This book takes a different approach. It seeks practical ways to minimize harm, prevent fraud, and avoid self-deception, ensuring that longer lives remain meaningful. In relationships, the answer comes with a set of habits. They won't make love simple, but they will help keep it healthy and alive.

Closing: Love as Constant Movement

Eli and Nora will start over many times before they decide their story is finished. They will argue and learn to apologize more clearly. They will change their minds and their way of living. In the fifth century, they will see new worlds in the sky and choose, once more, how to be together.

For the rest of us, we don't need to wish for endless time. We just need to learn from Eli and Nora's discipline. When forever is possible, the real skill is not holding on forever but choosing to renew. With the right structure, love becomes something you choose to do, gladly, again and again, through seasons that stretch across centuries.

Before longevity, our greatest issue was that we were running out of time. Every day, we had less time, which created the problem of constant hurry. We couldn't always hope for the best, because many times we were just taken by the stream. When we have immortality to enjoy, another problem arrives: We aren't pushed to decide, and when we need to do so, we're confronted by too many options to acknowledge and select. Immortality does not remove the burden of choice; it makes choice endless. And so, like Eli and Nora, we must learn not just to endure, but to renew.

Chapter 11

The Burden of Infinite Choice

What Changes When "Later" Stops Arriving

The sensation of urgency that arises when we perceive time as limited often compels us to prioritize, make decisions, and move past regrets. However, try to envision a scenario where time extends infinitely, such as an immortality project spanning entire centuries. The choices you make now cast long shadows, affecting future generations and landscapes that you yet can't even imagine. In such a scenario, questions of which changes to adopt would become pressing.

Our priorities would shift toward decisions that could be reversed, as even rare risks would eventually materialize. Besides being responsible for the world's children and their children, you would still be in that world. You would have to explain your choices to the people they go on to benefit or harm.

This means planning must encompass longevity, which presents many unknown challenges. If you have a short list of options, you might need to opt for something that isn't perfect. But what about when the options are unlimited? Searching for the perfect solution among many good options may lead to inertia.

Without natural turning points, your identity and values would risk becoming influenced by what is easiest in any given moment. To navigate this, you could begin by listing your current major decisions and classifying their reversibility. This concrete action would help ground abstract planning and offer a practical starting point for managing choices over an extended timeline.

Another idea to put this into practice is, before making every critical decision, to consider it in terms of how easily it can be reversed. Some choices are like two-way doors: You can go through, learn something, and come back without much trouble. Others are more like doors with stiff hinges: You can go back, but it costs you something, such as money, relationships, or your reputation. A few are one-way doors: Once you go through, you can't return, or going back would cause more harm than good.

We often get stuck when we treat easy-to-reverse choices as if they were final and spend too much time deciding. Big problems happen when we treat one-way doors as if we can undo them later. The key is simple to say but hard to do: Name the kind of door before you open it, and match your decision process to what you find.

Guiding Policy: Decide Often and Reverse Gracefully

The main idea is to make lots of small, low-risk decisions where you can easily and quickly change your mind. For choices that are harder to undo, break them down into steps so you can learn from them early and still have an exit. When a decision can't be reversed, take your time and set up ways to protect your future self and anyone who relies on you.

Recognizing Choices

Here are some questions and a checklist to help you recognize each type of decision in daily life:

1. Is this decision reversible?

   - Can I easily go back to my previous state or situation if I change my mind?

   - Will going back cause little to no consequence?

2. If it's reversible, what will I learn by going through this door?

   ○ Will this decision help me gain new knowledge or experience?

   ○ Is it worth trying even if I might go back?

3. If the decision is not easily reversible:

   ○ What will it cost me if I try to reverse it? (Consider money, relationships, reputation, time, and effort.)

   ○ Are the costs of reversing higher than the benefits?

4. Is this a one-way decision?

   ○ Once I commit, can I realistically undo it or return to how things were?

   ○ Would trying to reverse it do more harm than good?

5. How confident am I in making this decision?

   ○ Do I have enough information to proceed?

   ○ Have I considered the potential consequences of both going forward and going back?

6. What are the risks and opportunities of this decision?

   ○ What opportunities might I miss if I don't go through this door?

   ○ What risks am I exposed to by proceeding?

The checklist below can help categorize daily decisions into two-way doors (easy to reverse), doors with stiff hinges (costly to reverse), and one-way doors (irreversible). Ask yourself these questions

- What do I need to write down?

- How do I explore over long periods?

- When can I stop searching for the perfect answer?

- Which promises need regular check-ins?

- How do I notice changes in my values?

- Why is it essential to plan for mistakes?

- How do I think about the effects my choices have on shared resources?

The rest of this chapter will show you how to put these ideas into action.

A Docket for Consequential Decisions

Written words have power, particularly over the long term. Writing things down isn't just a formality; it's a valuable tool. A first helpful strategy to master decision-making forever is to make notes. Here's a simple step-by-step guideline of how and why:

1. For any big or hard-to-reverse decisions, make a one-page summary before you act. Start by naming the real problem you're solving and what success would look like in five and fifty years.

2. List the options you thought about and why you didn't choose them, even if the reason is simple.

3. Say how easy it is to undo the decision and explain your thinking.

4. If it's difficult or impossible to reverse, consider adding safeguards such as a waiting period, clear rules for backing out, and a trusted person who can step in if needed.

5. Note any backup plans or exit options you're keeping open. Decide how much failure you can accept, how you'll spot problems, and who will watch for them.

6. Be clear about which resources and people your decision affects, and check your plan against public standards if it's a big one.

7. Finally, set a time to review the decision and name someone to handle it, because if no one is responsible, it's easy for things to be forgotten.

The goal of this document isn't to look serious but to make your thinking clear in a way that your future self and others can understand. If you want your family, colleagues, or even a court to take your commitments seriously, they'll need more than your word; they'll need something written that lasts beyond your changing moods.

Exploration and Exploitation Across Centuries

Within organizations and AI systems, balancing the pursuit of new opportunities with maximizing current benefits is a well-recognized challenge. Investigate to identify valuable options; capitalize on them to gain rewards. For mortal lives, the exploration window is short, but for immortal lives, "early" is long but not infinite, while "late" can arrive surprisingly quickly when your attention is stuck in comforts.

A workable flow is to overweight exploration in the first century:

- Spend most of your time sampling roles, geographies, disciplines, collaborators, and ways of living.

- Keep your commitments short and reversible, and write your experiments as experiments so your identity does not overfit to the first interesting result.

- Consider the career trajectory of a data scientist: They take on multiple short-term projects across various industries such as health care, finance, and technology, gaining diverse insights without binding themselves permanently to any single path.

In the second phase, which you can think of as the next two centuries, adopt something like a barbell:

- Protect a resilient core portfolio of roles and relationships you intend to keep across eras.

- Make regular risky bets in a specific area designed to stop things from becoming too fixed or unchanging.

For example, the data scientist might choose to anchor themselves at a research institute, developing deep expertise in AI ethics, while also collaborating on high-risk, innovative start-ups in areas such as quantum computing.

In a third phase, focus on areas where your efforts build value over time, like basic research, long-lasting organizations, and communities you manage, while keeping a small, ongoing investment fund that stays open. As the data scientist matures, they may choose to focus on establishing a leading think tank dedicated to AI issues related to public policy, ensuring their work has a long-lasting impact. Meanwhile, they can still keep a few exploratory projects going to maintain a flow of fresh ideas.

Researchers have addressed exploration and exploitation and highlighted risks and challenges.[1] They warn that we should not idolize novelty at the expense of compounding value. Life is not a calculation where we should always choose the immediate best option. It allows room to accept satisfactory outcomes while exploring unfamiliar opportunities.

Real Options and Staged Commitments

Think of your big choices as options with staged exercises, not single, gigantic bets. In finance and in product development, real options give you the upside of learning while capping the downside. In personal and civic decision-making for immortals, the same logic would save lives and years.

A staged exercise is a step-by-step activity designed to guide someone through a process or concept in a gradual, structured way. Each stage builds on the previous one, helping participants understand and apply ideas progressively.

For the decision recognition checklist, a staged exercise could look like this:

- **Stage 1: Identify decisions:**

  - List three recent decisions you've made in your daily life.

- **Stage 2: Classify each decision:**

  - For each decision, answer: Was it easy to reverse? Did it cost you anything to reverse? Was it irreversible?

- **Stage 3: Reflect on the consequences:**

  - For each decision, write down what you learned or what the outcome was.

- **Stage 4: Carry out a decision impact assessment:**

  - Assess if any decision was more costly to reverse than anticipated or if any was truly one-way.

- **Stage 5: Apply awareness:**

  - Think about an upcoming decision and use the checklist from earlier in the chapter to identify what type it might be before deciding.

This gradual, hands-on approach helps deepen your understanding by doing, reflecting, and planning. Maintain credible alternatives in parallel for long enough to keep your main path honest; watching a second plant take root keeps you from pretending your only plant is thriving when it is not. Refresh your health options on a schedule; this ensures you don't carry zombie options that consume your attention because you are nostalgic for who you were when you first opened them.

This staging logic is particularly challenging when your choice involves public resources and private grids. Think about, for instance, a lab you intend to scale, a continuation service you plan to open to others, or a campus that will alter land and water. The rails and limits from earlier chapters show the practical constraints that frame and influence decisions, such as clear energy and water budgets. These must be considered and exposed so everyone understands your choices.

Satisficing and Where to Optimize

When you have endless options, trying to make the best choice every time can leave you feeling stuck and unhappy. Studies show that having too many choices leads to less satisfaction.[2] In the long run, it's better to decide what's "good enough" for things that aren't high-stakes, and then pick the first option that fits. Save your energy for the few things that truly matter, such as keeping promises, caring for the environment, or doing work that shapes who you are. Most everyday choices, like what to buy or where to travel, don't need that much attention.

One of human beings' hardest burdens is the fear of regret and the experience of regret. When adding years and years to our lifespan, the risk of feeling regretful about how things turned out significantly increases. Over centuries, counterfactuals would proliferate until they feel like ghosts.

In an immortal life, regret could feel endless because you would have infinite time to revisit past decisions and their consequences. Yet, this unique condition would also offer a profound opportunity: Unlike in a finite life, you would have countless moments ahead to learn, adapt, and grow from every choice. Regret, then, would become less of a burden and more of a guidepost, signaling areas for understanding and improvement. Here's a simple guideline that can help:

1. Acknowledge regret without judgment:

   ○ Accept your feelings of regret as natural and valid.

   ○ Avoid harsh self-criticism; instead, see regret as information.

2. Reflect on the lessons:

   ○ Ask yourself what this regret teaches you about your values, desires, or decision-making.

- Consider how this insight can inform future choices.

3. Reframe regret as growth:

- View regret as a step in your continuous personal development.

- Recognize that even immortal beings evolve through their experiences.

4. Experiment and take new paths:

- Use your endless time to explore alternatives and new experiences.

- Don't be afraid to take risks, knowing you can learn and adapt indefinitely.

5. Practice compassion and forgiveness:

- Be kind to yourself for past choices.

- Forgive yourself as you would forgive others for mistakes.

6. Focus on the present and the future:

- Ground yourself in the here and now.

- Plan and pursue meaningful goals, knowing each moment adds value.

Remember, immortality doesn't mean perfection; it means an ongoing journey where every experience, including regret, can contribute to your wisdom and resilience.

Precommitment and Guardrails

Precommitment devices exist to protect you from you. They trade autonomy now for resilience later. The simplest and most humane method is a cooling-off period between intent and action for any choice you classify as one-way. You will save yourself from choices made under sleep debt, grief, or anger by enforcing a delay for you to take a pause and ponder your options. It's like a veto power on yourself.

Consider a set of high-scope moves: public continuity claims, guardianships, and irreversible identity edits. For these, you might want to implement the following measures:

- Constitute a small review council with a mix of long-time intimates and neutral expertise.

- Constrain its mandate tightly so it does not become a fixed regulation in your life.

- Rotate its membership so it does not remain unchanged and outdated.

- Bind renewability to roles and control.

As you saw in the governance chapters, time-to-live limits and reauthorization windows keep institutions alive, so you should adopt the same techniques privately.

Value Drift and Identity Governance

You will change. The question is whether the changes are coherent and understandable to the people who rely on you and to your own sense of self. Treat value drift like any other kind of drift you would monitor in a control system. Some aspects to consider are as follows:

- Maintain a short baseline statement of why you do what you do. Every 25 or 50 years, review that baseline and write a note.

- Record the trade-offs you've made that would have surprised your earlier self, and explain why you think they were right.

- When the drift alters obligations you took on under different values, share your notes with the people whose lives your choices underwrite, like your family, colleagues, and communities.

- For projects and commitments involving public beneficiaries, embed drift into governance to ensure that your changes do not disadvantage others.

- Continuity trusts and branch-aware controls exist to prevent one person's evolution from becoming another person's damage. In civic life, keep the parity rule tight: one identity, one civic vote, even if you have branches.

It helps to admit what is philosophically obvious and emotionally challenging: Continuity over long spans is partial, not perfect. You should be humble and check the facts carefully when you say something that connects to the past. Be careful not to assume that what you believe now is always true about what you promised before. It also helps to remember Scheffler's reminder that most of the value you help create will ripen after you. Even immortals are parts of projects larger than any single consciousness.[3]

Heavy-Tail Risk and Error Budgets

Over eternity, rare events are not so rare. An error rate that appears irrelevant on a one-year horizon becomes a major flaw over 10,000 years. Consider a simple probability rollover: A 0.01% yearly error rate translates to a 63% chance of occurrence over 10,000 years. Such numbers emphasize the inevitability and impact of what initially seems unlikely and demand a deeper commitment to prepare ourselves for all eventualities.

Systems that work fine under average loads fail under stress regimes that you do not experience often... until you do. This implies adopting error budgets and postmortems for both personal and institutional decisions. Ways to be prepared include:

- For each system you run, including labs, clinics, communities, and continuation services, define a tolerable failure rate and the conditions that trigger a rollback or a freeze on new enrollments.

- Practice outages and drill black-starts.

- Publish incidents to the populations they affect and to your own archives. This is an ethical practice that keeps planes in the air and grids from cascading.

You will be tempted to believe you can fix your way out of every failure. Sometimes you can; sometimes you must stop and change what you are doing. Red-team your own plans before scaling choices that you cannot easily reverse. If you cannot, find people who will tell you that you are wrong and who can bring fresh ideas to build new solutions.

Externalities and Stewardship

Your choices will run on electricity, water, and the patience of communities. Treat those facts as first-order variables, not afterthoughts. If your life or work depends on significant computing, you are part of an energy system that must be clean, reliable, and accountable at the same time. When you scale continuation services for others, tie your compute spend to biodiversity funding through the tithe, and publish your grants and results. If you are opening a campus or cluster, publish access metrics by geography and income, and adjust budgets as needed to address persistent gaps.

From Rules to Practice: Three Extended Cases

A framework is only as good as the lives it can carry. The three cases below illustrate how the rails translate when they meet the weather.

First Scenario

First, consider a three-century vocation arc: an engineer in their first fifty years samples roles across human–machine interfaces. The decisions are reversible by design. The engineer visits the researcher's post for two years. They hold partnerships that can be unknotted, and the pilots are small enough to fail without ruining relationships. A docket exists, but it is simple and serves as a memory aid more than a legal instrument.

In the second century, the proofs are growing, and the engineer is better at creating systems to check materials for safety than at inventing new materials. Decisions now focus on flexible options. Starting a lab costs a lot and cannot be undone, because closing one causes problems, hurts careers, and wastes money. Meanwhile, more tasks pile up. The engineer knows they need clear steps before growing the lab; problems that might stop work are planned for; a limit on mistakes is set. They show energy use and data sources clearly on public websites. Rules for getting permission are shared publicly. A fee to keep the lab running is included in contracts.

In the third century, the lab's methods are the national baseline. A one-way door appears: a claim of personal digital continuity as the "same someone" for narrow legal purposes. The docket now can't be changed because it is the mainframe. Cooling-off is set at a year, and the teams display all the verification regimes. The claim is carefully bounded: It is valid for closing a specific promise under consent and audit, but not for corporate control or political representation.

In the next fifty years, a different one-way door appears: whether to scale the lab into a global provider under the founder's control. The answer is no. The control rights are vested in a continuity trust with terms governing leadership, sortition, and reauthorization control. These methods are placed within an open consortium. The dashboards show access widening in regions that were previously locked out; the tithe funds support wetlands and the clinics on which the lab's work depends.

Second Scenario

Second, let's imagine a values-edit proposal. A scientist notices that ap-
proval-seeking behavior on social platforms erodes attention for long-horizon
work. They consider a targeted edit to dampen responsiveness to social reward
signals.

The choice is one-way. The manual prioritizes staging. They begin with
training and environment redesign, proceed to pharmacological modulation if
necessary, and only then consider an edit to gene expression. Cooling-off is six
months, and a future-self individual has veto power if sleep debt or bereavement
is present. A small guardian council includes a clinician and a friend who has
disagreed with the scientist in the past and is trusted exactly for that reason.

They have developed an error budget. If measured empathy drops more than
one standard deviation from baseline, the experiment stops, and they attempt
a rollback where possible. The workload is reduced while the system returns to
equilibrium. As it happens, training and the environment redesign the work,
and they never attempt the edit. In the end, they archive the manual, update
the notes to register the changes, and measure the outcomes.

Third Scenario

Third, think about a city-scale retrofit program. The founders aim to decarbonize buildings, reduce heat exposure, and enhance indoor air quality. The
architecture of the program is irreversible in terms of cost: Once structures have
been bonded and the authorities have been established, they are difficult to
reverse. The manual treats the one-way elements as public-goods decisions from
the start.

Extraordinary powers sunset by default. Citizens' assemblies chosen by
stratified sortition review recommend significant amendments on a published
timetable. Time-to-live limits roll leaders off before incumbency turns into

ossification. The program's dashboards report retrofit rates, equity metrics by neighborhood and income, grid reliability, and embodied carbon.

A reliability drill fails in year 34, and the system expects the incident report in three weeks. The teams shift the budgets to shore up the weak points and pause expansion until the drill is passed. At the reauthorization window, the program narrows the powers it no longer needs and renews those it does with better constraints.

Standard Failure Modes and What to Do About Them

Perfection paralysis is the quiet killer of long lives. The cure is to respect your own classification. If the decision is a two-way door, cap preparation, set a deadline, and go. The point is not to get it "right" but to learn cheaply.

Continuing to invest in a project that is already lost is a common problem. The best way to avoid this is to set clear review points where you stop unless there is strong proof to keep going. Give someone you trust the power to stop the project if needed. Identity myopia happens when you make a one-way decision based on a temporary state of mind due to a lack of sleep, sadness, jealousy, or fear.

The solution is simple:

- Set a rest period you cannot skip.

- Rely on a trusted person, whom you picked when you were clear-headed, who can stop decisions.

- Establish and follow a clear rule against making permanent changes under pressure.

People often ignore the hidden costs in rich groups: You use more compute power, water, and attention because the bills don't come right away. The important thing is to make those costs clear and real by sharing public reports, investigating problems, and stopping growth when the numbers do not support it.

Changing values without keeping records breaks trust. The way to protect trust is to record changes, keep track of agreements and time limits, and accept that some projects should end with a public note and thanks rather than continue causing damage.

A Note on Ambiguity and on What to Count

If this chapter seems light on philosophical jargon, that's intentional, but the ideas are simplified. *Reasons and Persons* by Derek Parfit (1984) is an in-depth exploration of personal identity, ethics, and rationality for the kind of continuity that matters for ethics and law.

Parfit challenges traditional views about what matters in survival and personal identity, proposing that identity is not what fundamentally matters in survival. Instead, psychological continuity and connectedness are key. The book also addresses moral philosophy, including how we should make decisions that affect our future selves and others, and scrutinizes common assumptions about rationality and self-interest.

Prospect theory gives a more accurate picture of how you'll actually feel about losses and gains compared to a lot of idealistic talk about "rationality".[4] Meanwhile, research on the paradox of choice explains why having too many great options can leave you feeling worse than just having a few.[5]

Long-termist thinking helps you understand why it's important to protect future possibilities, both for yourself and for others, as a responsibility rather than something to avoid.[6] Bostrom's work on existential risks is crucial because your choices can make these risks bigger or smaller; there's no room for mistakes when the consequences are permanent.[7]

Scheffler and similar thinkers remind us that meaning, even for those who live a very long time, comes from being part of projects that outlast a single life and are finished by others.[8] You can use their ideas without having to come up with everything from scratch.

Closing: Ethics as Flow

Having endless time doesn't mean you can act without care. It means making choices that keep your life your own, keeping your promises clear, and not putting hidden burdens on others. The main advice is to sort your decisions before making them. Settle for "good enough" when there are too many options, and focus on getting things right where mistakes would really matter.

If you keep this rhythm, "forever" stops feeling overwhelming and becomes what it should be: a long stretch of tomorrows, each chosen on purpose, each reversible when possible, each anchored when needed, and each lighter because you've set up supports to carry the load.

So far, we've explored how our known world might adapt to a new reality in which lifespans doesn't have a biological determinant. However, most of this is just an update to what we already use. Now, it's time to go further and enter the world of imagined yet possible alternative worlds. What will life and the world look like in a post-human experience?

Part VI: The Varieties of Post-Human Experience—Exploring Different Paths to Transcendence

Chapter 12

Virtual Worlds as Primary Reality

The Civic Weight of a Digital World

On a wet April night, Nia sets her tea down, puts on soft cuffs that feel warm, and gestures toward the glowing orb showing her father's usual face. Three generations live on their block, and all three inhabit a shared city woven from memory, craft, and code.

Nia is a hybrid: She cooks and bikes and laughs in a kitchen that smells like cumin and soap. She also teaches and builds in a lattice of rooms that opens like a book when she closes her eyes.

At the corner of her street, a bulletin blossoms: The platform is changing the conflict-resolution rules in public squares. The last time an update shipped without warning, her grandmother's memorial garden flickered for hours and the lilies lost their scent until a rollback patched the error.

Nia taps the bulletin and exhales. A docket opens with signatures, a change log, energy and water disclosures for the compute expansion that will support the new mediation system, and a drill date. A citizens' assembly by stratified sortition will review the change; "one identity, one civic vote" will assure parity when the city ratifies it.

She smiles at the mundanity of it all. *This*, she thinks, *is what a world feels like when it is virtual: You can change the physics, but you still cannot skip the rules.*

Diagnosis: Why Virtual Primacy Changes the Problem

Virtual environments were once on the sidelines, serving as playgrounds, training spaces, or labs for experimental art. Now, as people spend more time in these environments and as social systems adapt, these spaces are becoming central to work, education, healthcare, art, civic life, and family traditions. This shift is changing every aspect of social life, resource allocation and expenditure, and even how people find meaning and face challenges.

The first change is cognitive and emotional. Programmable environments can tailor challenge, pace, and feedback in ways physical spaces cannot. Good designs meet people where they are and nudge them forward toward competence. The literature on flow and self-determination has emphasized for decades that mastery appears when challenge matches skill and when autonomy, competence, and relatedness are respected.[1]

For example, a challenge-to-skill ratio of approximately 4:3 is often cited as optimal for maintaining engagement without overwhelming the participant. Bad designs, however, are just as powerful but in the other direction. Easy personalization can remove the small challenges that help people learn. Focusing too much on keeping attention can create habit loops that tire people instead of helping them.[2]

It's natural to want to turn unfamiliar things into something easy and known, especially when life feels like it's going your way. But a good life comes from balancing who you are with the challenges you face.

The second change is infrastructural and ethical. When livelihoods, appointments, and continuations depend on platforms, "uptime" becomes a moral obligation. This has two implications: Reliability becomes a first-order design criterion, while accountability means following clear steps: running tests to check if power systems can restart on their own and work independently during problems, sharing reports with the public when things go wrong, and making sure licenses depend on how reliable the service is. Treating power outages like small technical problems is wrong when important places like clinics and schools depend on those systems.

The third change is political. Virtual primacy concentrates unprecedented power over attention. Design choices that prioritize engagement over reach can lead to systemic inequalities. Some voices become inescapable, while others are never surfaced. Society can fall victim to the tyranny of the algorithms, with civic agendas following popularity rather than deliberation. Without firewalls, popularity in cultural spaces seeps into civic influence and corrodes the concept of "one identity, one civic vote."

The fourth change is ecological and material. The lightness of pixels can be misleading, but the servers that keep parliaments, parks, and classrooms alive run on electricity and water and are made of materials with an embodied carbon footprint. Location matters. Cooling choices change river temperatures. "Clean" power on an annual contract can hide hourly dirtiness if 24/7 matching is not enforced.

A final issue, on a philosophical yet practical level, is related to identity and legitimacy. In virtual polities, one person can appear as many, and synthetic agents can mimic living. Continuations can reconnect families or exploit them. Without verification regimes, provenance chains, parity rules, and renewal rituals, you cannot make fair claims about who is present or whose voice counts.

Guiding Policy: Treat Virtual Primacy as Infrastructure and Polity, Governed by Rails

Wherever a capability becomes a substrate for human life, it must be governed by rails that make it reliable, legitimate, and fair. If we apply this approach to virtual primacy, the guiding policy crystallizes.

Verification and Provenance

Where an agent claims continuity or authority, those claims must be supported by auditable tests and signed chains of evidence from data to model to deployment. World-logic updates, feature flags, and policy changes should have manifests, diffs, and rollback plans, while emergency fixes should be followed by postmortems that provide clear reasons.

Consent

The more intimate the intervention, such as merges, memory-sharing, or telemetry that can be reassembled into inner life, the more you need periodic renewal. This renewal must be under precise terms, with revocation that actually propagates to derived artifacts.

Equity

Baseline access to compute, storage, and acceptable latency windows must be a right, not a perk. Public dashboards should display access, safety, and harm data disaggregated by geography, income, and race/ethnicity. The governance rails must be viewed as inevitable foundations rather than optional enhancements. With rails, we ensure stability, transparency, and collective accountability, akin to the triumph of the open standards that shaped the evolution of the Web. Without them, we risk a fragmented and inequitable landscape.

However, it is essential to acknowledge potential criticisms of this approach. Some may argue that such comprehensive governance could stifle innovation or lead to significant bureaucratic overheads that outweigh the benefits. Others might question the feasibility of implementing these governance rails across diverse virtual environments with varying needs and resources. We need to address these objections to make sure that governance frameworks are flexible, scalable, and adaptive, capable of evolving as technology and societal demands change.

These rules don't limit creativity; on the contrary, they protect it. Without them, art, learning, and play can be undermined by fragility, fraud, unfairness, and a loss of trust. With them, a city like Nia's can grow quickly while still supporting the people who rely on it.

Designing Worlds That Stay Meaningful

Because virtual worlds can give people whatever they ask for, designers must resist the impulse to treat satisfaction as a simple function of fast gratification. Flow persists only when systems maintain an edge between mastery and mystery. This means creating situations with challenges that are carefully adjusted over time. Others can offer support at first but gradually remove that support as skills improve. It also means offering new experiences regularly without just adding new things for no reason.

Self-determination theory can help us understand this. Autonomy requires genuine choices that alter outcomes and tools that enable people to author their own paths, not just select from menus. Skills grow best when feedback is clear and helpful rather than just offering praise. Getting a badge for showing up is very different from improving in a skill. People need places where they can feel seen and appreciated. This means using tools that give credit fairly, allow sharing with permission, and have rules that recognize mentoring, translating, and moderating as valuable work.[3]

Co-creation should be the default. Shared editors, version trails, "fork and merge" capabilities for worlds and artifacts, and remix affordances with clear provenance push culture beyond "consume and comment" into "make and maintain." Different types of hospitality help build trust. Clear, flexible spaces allow people to be creative, while quiet, private areas give a stable place for those who need it. There would be clear documents explaining the rules and permissions of what is polite in each area.

Moderation is the system that keeps things running smoothly. It needs trained staff, clear rules, enough time, and ways to measure how well it works.

People will track problems with trust and safety, like technical problems, and fix them regularly, not just during a crisis.

Cultural continuity goes beyond nostalgia. It is an agreement to remember together. Rituals and festivals are meant to renew. Examples could include days when neighborhoods rewind to earlier states, exhibits of the "before" of a frequently revised room, and anniversaries that include revision rather than only commemoration. All these teach communities how to survive change without losing themselves. Archives that hold both artifacts and their diffs would make history a living resource rather than a museum.

In all of this, the goal is to keep virtual primacy from becoming private perfection. A world is something more than a series of personalized stories. It needs shared rooms, shared rules, and shared memory.

Measurement: Count What Counts and Publish It

Let's now focus on what and how to measure to keep life honest. Sharing information can lead to violating privacy, so some measures must be considered:

- Show baseline compute allocations and latency windows by neighborhood and income.

- Show usage and wait times so residents can see whether queues are fair.

- Publish harm incident rates and time-to-help metrics for harassment and abuse cases.

- Reliability should be staged and measured, and drills should have pass/fail criteria.

- The mean time to recovery from outages should be reported.

- Degradation under stress should be described in everyday language.

In one neighborhood, for example, public dashboard data might show high solo-uptime ratios, sparking a community meeting that leads to shared events designed to improve engagement. It would illustrate the effectiveness of Jacob's idea of "eyes on the street" applied to the virtual realm. Everyone would be informed, and would be able to make improvement suggestions and contribute to their community.[4]

To ensure equity, cities would need disaggregation in the metrics displayed on the dashboards. Mental health and engagement could be assessed with respect to defaults rather than attempting to psychoanalyze users. Some relevant questions to consider are as follows:

- What fraction of users opt out of time budgets and quiet modes?

- Are binge-friction patterns reducing overnight exposure without depressing satisfaction or learning?

- Are creator care programs reducing churn and burnout?

Another aspect of equity is respect for diversity. These metrics measure how often users encounter unfamiliar creators and topics and how many recommendations surface the unseen.

Risks and Mitigations: A Failure Log You Should Expect and Survive

Virtual extreme skepticism is the risk that personalization and safety turn into isolation. Its early signs are as follows:

- The time people spend in shared spaces shrinks.

- Their willingness to tolerate disagreement falls.

- Their narratives emphasize the inferiority of others' contributions.

Societies will develop mitigation strategies that begin with time budgets, quiet times, and "reality calibration" sessions. In these sessions, personalization is deliberately suspended, allowing people to expose themselves to a wider world.

It continues with architecture. Some features can require participation in the community to unlock:[5]

- Periodic "handshakes" keep a person's membership in a commons alive.

- Transparency alerts nudge people to notice their solo-uptime ratios.

- Culture-mentoring and co-authoring are measured as valued contributions.

Reliability failures are inevitable in complex systems, but they turn into problems when they cascade into harm. The indicators of fragility are failure without consequences, a mean time to recovery that stretches, and silence about data loss and degradation. The solution always takes us back to public, reliable information to ensure accountability and timely action.

Harassment and toxicity are as old as crowds. They become systemic failures when reporting is slow, enforcement is inconsistent, and creator care is an afterthought. Funding moderation, measuring response and satisfaction, and tracking enforcement consistency are the basics, as well as mental health defaults and creator care. The public should be able to see that a safety apparatus exists and that it is continually improving.

Identity drift without clear rules happens when promises made based on one set of values are later ignored without explanation. Regular updates on value changes serve as honest notes to future selves and team members about why priorities changed. Time limits on control rights stop one person's life story from controlling an organization forever.

Fraud and impersonation are the dark twin of continuity. Their early signs are reports that synthetic agents feel "off" to intimates. The practice requires implementing rigorous verification protocols and authenticated update sequences before accepting acknowledgments. All claims must remain tied to their originating context even after validation, and there must be strict penalties against identity fraud to prevent exploitation and harm.

A Worked Example: Five Years of Astra Commons

Consider a midsize metropolitan region that has launched Astra Commons, a public-option "city layer" intended to make work, school, care, and culture available in shared virtual rooms with the same seriousness that streets and libraries carry in the physical city. Astra Commons follows all the expected rails: identity governance, the principle of "one identity, one civic vote," and a branch registry that establishes the genealogies of selves for contracts and liabilities. Citizens can see all the updates on a public dashboard.

In the first year, the city sets up the identity registry and the compute commons. Platform contracts require 24/7 clean power matching for expansions, so the siting of a new data hall near a stressed river is halted until an air-side cooling redesign lowers return temperatures. The first black-start drill fails, which the city makes public in a two-week after-action report. Budgets shift to address redundant switchgear, and the government holds further expansion until a second drill passes.

In the second year, a stratified citizens' assembly reviews an attention firewall proposal. They reflect on the diversity indicators alongside engagement numbers. They decide to expand creator care funds, reduce burnout, and improve harassment response times. The continuation compute tithe flows to a regional watershed restoration program, and the city publishes the names of recipients and their five-year goals.

In the third year, the metrics show that solo-uptime ratios are rising in two neighborhoods with a high concentration of older residents. These people migrated into Astra Commons after the closure of physical community centers. The city launches "reality calibration" sessions that are funded and facilitated as part of public health. The community adds shared challenges that open new gardens and studios, and the regions' solo-uptime ratios fall without depressing satisfaction.

In the fourth year, a small change in world law accidentally reduces the rights of people reporting abuse. The change was reversed within an hour. A public report explained what went wrong and how it was fixed. A new rule says that reliability tests and drills must be done for each feature, not just for the overall system. Two companies lose the right to grow until they prove they have passed these tests with public reports.

By the fifth year, access disparities have narrowed meaningfully. A report on carbon emissions from a planned expansion sets a stricter carbon limit for the next year to keep the region within its goals. More people agree to participate in random selection processes because financial help and childcare support make it easier. The city releases a "state of the world" report that is straightforward, showing successes, problems, numbers, and future plans.

Decision Architecture for Users and Builders

Individuals and institutions alike benefit from classifying choices by their reversibility and writing them down before taking action. People can reverse decisions with minimal effort, which warrants swift trials, clear metrics, and rapid implementation. They can launch a new creative guild, tweak default lighting settings, or participate in a collaborative teaching group. Choices that require significant resources or consequences to change, such as altering a moderation system on a major platform or increasing instructional hours for young students, demand thorough planning and more detailed procedures. Some of those choices

One-way doors deserve cooling-off periods, veto powers, and public journals. Large-scale identity edits and world laws that alter civic rights should be addressed here.

A one-page docket is enough to prevent self-deception while maintaining momentum. It sets purpose and success signals in the medium and long term. It lists the alternatives considered and why they were rejected; even "I was exhausted" can be an honest reason. It explains the classification, the externalities, and other rules related to consent.

Closing: Virtual Primacy That Deserves the Name "World"

The real promise of virtual primacy isn't about erasing inconvenience. It's about giving people more control over their lives, communities, and institutions, with stronger safeguards in place. The best virtual worlds help people find focus without addiction, connect without pressure, and create without burning out. They remember together, practice, and aren't afraid to make changes and try again.

If we treat virtual primacy as a public utility and a democracy, the best parts will shine and endure. What sets a true world apart are the promises it holds up through problems, changes, and time. How do we achieve that if we keep seeking perfection?

Chapter 13

The Physical Perfectionists

To Enhance Without Erasing

On clear mornings, Anika runs the river path before her clinic opens. Between each stride and each breath, she embodies her core promise: to enhance without erasing, ensuring that each advancement respects the natural rhythm of life. A mist rises off the water where the restoration team reconnected a cut-off oxbow the year before; egrets fish in the shallows.

In her vest pocket sits a consent packet for the week's partial reprogramming trial. This involves giving targeted doses to specific tissues for a short period to study particular effects, with a clear plan for stopping and observing over time. That is the longitudinal monitoring. The consent will renew this evening, as it does each year. At the clinic last month, she and her staff conducted a black-start exercise with the water utility and the small data center that hosts their registries. The drill report is public.

Between her breath and the river lives the reason she chose a biological path: to enhance without erasing. When she hands her patients tea after consultations and reviews the registry of the real reductions in hospitalization days, the adverse events logged and trended down, she thinks about it similarly to her runs: Go further when it is safe. Stop when it is not. Keep the promise to the body and to the place.

Diagnosis: Why a Biological Path Endures

A primarily biological path lasts because our bodies shape how we think and feel. Signals like heat, breath, and the ache after effort are sensations that help us understand what matters. People who choose this path want to maintain that feedback as their normal state, rather than relying on constant outside mediation. They accept helpful tools and care, but they do not want to change their basic way of living.

Even if, in the near future, some families accept a carefully verified claim of digital continuity, the individual needs to meet a list of requisites:

- tests for behavioral concordance under novelty

- autobiographical recall with context

- trait stability over time and learning

- relational validation by people with extended memory

However, informed people who are ready for digital continuation will still disagree about subjective continuity.

Imagine a scenario where two siblings face a decision regarding their aging parent, who is considering a digital transition. One sibling, captivated by technology, sees digital continuity as an opportunity to preserve their parent's essence and wisdom. The other, deeply rooted in the natural course of life, is dubious about whether the digital representation could capture the true spirit and history embedded in their parent's lived experiences.

This divergence is a good example of the complexity of subjective continuity. Simply put, people will have different opinions about whether a digital version of a person truly maintains the same essence as the original, living person. Some prefer to stay with the biological organism that has carried a life's promises so far without needing metaphysical agreement. This approach helps prevent harm by avoiding a simple yes-or-no decision when facing deep uncertainty at life's critical moments.

Another reason is that the biological path has a different set of risks and relies on medical systems people are already familiar with. Some people will trust advanced neural interfaces and virtual-first approaches. Others will opt for biological interventions, which carry risks and require more oversight, similar to traditional medicine, rather than software. They are willing to follow careful tracking, reporting of side effects, and changes in training as signs that these methods are truly being adopted. They seek secure, relevant outcomes.

There is also an idea about identity behind the practical actions: Living a natural life goes well with caring for the environment. The same attention we give to healing and caring for our bodies applies to rivers, soil, and trees. A society that wants to inherit a safe world considers caring for the body and taking care of the environment as part of the same path.

We must be honest about the costs involved. Biological enhancement is not free of charge, and we must track the real side effects. We must also consider the proactive steps to ensure access. Consider a typical individual named Sam, who lives in a rural area and earns a modest income. She is interested in biological enhancements to help manage a chronic health condition. However, the high costs and limited access to nearby facilities mean she has to travel long distances and save up for each appointment. Even the most natural approaches depend on equipment, sensors, and supply chains.

Guiding Policy: Enhance Without Erasing

The main policy prescribes making improvements within the body in verifiable ways. The first rule is to focus on decisions that can be reversed, plan carefully for those that are harder to change, and take your time with decisions that cannot be undone.

The criteria when considering, checking, and approving an improvement are to count only what moves hard outcomes across years. To make sure the criteria are properly applied, you need preregistration protocols linked to third-party audits. The different stages of a process must be recorded so a family, a clinician, or a regulator can see not only the result but how it was achieved. We must aim to avoid spectacular before-and-after images or biomarker shifts to stand in for event-level reductions in disability and death.

Ultimately, taking care of the world is a responsibility, not an added burden. The benefits people derive from advanced biological services should be linked to efforts that restore nature and protect water through a small, transparent fee that supports conservation trusts. Information about who receives these funds, where the work takes place, and what results are achieved must be in the public domain. Throughout, remember that good management and fairness matter: Success in one area does not mean someone should have more votes or special access. Everyone still gets one vote, and organizations should not mistake popularity for absolute authority.

Coherent Actions: From Principle to Practice

A practical approach to biological care uses the same structure that we have already outlined. It combines various tools to manage risk and connects benefits to environmental care.

Implementation begins with the interventions themselves, ordered from risk modification through repair and monitoring. Somatic gene therapy targeting common risk factors such as lipid handling, coagulation thresholds, and amyloid processing is an essential tool when monitored over a long period. It helps by reducing the chances of acute health events, such as heart attacks and strokes, thereby extending healthy living years and improving quality of life.

Cellular senescence modulation brings benefits, particularly for conditions where senescent cells contribute to inflammation and tissue damage, such as fibrosis or certain metabolic diseases. This intervention can improve quality of life by managing these conditions more effectively.

On the other hand, translation for general longevity use demands caution. To ensure safety, medical teams must commit to implementing intermittent dosing, careful cohort selection, and a registry that accurately captures adverse events. Partial reprogramming offers the potential to reverse specific functional losses when applied to tissue, but it requires careful control to avoid dangerous growths that could unfold other damages. Regeneration using autologous cells and decellularized scaffolds is transitioning from high-end to accessible care in specific treatments, with a focus on durability and ensuring broader access for these advancements.

Instead of focusing on intricate details like cytokine flares, arrhythmic precursors, and clotting shifts, imagine continuous monitoring as a vigilant watchtower, spotting early signs of potential health issues and alerting medical teams accordingly. These sensors are appropriately managed, operating on clean energy, and, should the power falter, they remain ready to handle the unexpected. Additionally, they ensure that all collected data remains trustworthy, with privacy safeguards to protect individual dignity.

The emphasis on monitoring processes and outcomes reaches regular renewal of consent. During renewal, the doctor explains results and side effects in simple terms, points out new options, and allows people to pause if they are unsure. For choices that cannot be undone, there are waiting periods, and a small group is available to step in if needed. All decisions, options, risks, and exit strategies are documented.

Impact on Culture and Sports

Culture, sport, and art as human activities will change as bodies and minds do. Category frameworks that distinguish between unmodified, medically necessary enhancements and declared voluntary enhancements would allow competition and performance to flourish without pretending that differences do not exist. However, these aspects would need clear safety standards and provenance requirements. Institutions must work on protocols that pre-establish doses, devices, and adverse events. For elite programs, especially those that benefit from public attention and funding, these protocols must carry inclusion obligations to open clinics, share methods, mentor, and reserve training seats for underrepresented participants.

Records should be categorized under choices that match the selections made. No category can be seen as morally superior. The point is to acknowledge differences and set healthy ways to balance them instead of denying them.

Preserved Natural Environment

There must be a plan to preserve nature regardless of the humans who choose to live in it, whether they choose continuation or not. This plan should aim to protect and connect ecologically representative land and sea, restore degraded systems, and measure survival and function. Indigenous peoples and local communities would play a key role, as we saw in previous chapters.

Clinics could contribute a bio-stewardship tithe to Perpetual Conservation Trusts. Sensor networks that monitor watersheds would disclose their own energy and water footprints and undergo drills, just like any other critical platform. They would also establish restoration projects with survival metrics and maintenance budgets.

Let's consider a real ecological emergency. A valley's streams, once teeming with trout, are now struggling in the increasing heat, pushing these fish to retreat to the cooler, higher tributaries. Under the weight of this loss, a coalition emerges out of necessity. This group, comprising an Indigenous council, a farm bureau, several clinics led by biological purists, and a regional parks agency, commits to a decade-long coordinated response. Their strategy includes widening riparian buffers, re-meandering a straightened reach, and adding large woody debris to enhance habitat complexity.

They build fish passages at previously impassable culverts, and a community nursery engages teenagers in cultivating native willows. As a result, summer stream temperatures demonstrate cooling trends, and flood risks diminish as waters are managed more effectively across floodplains. Meanwhile, clinics contribute by transparently disclosing their energy and water usage, dedicating a portion of their revenue to corridor acquisitions and monitoring. This represents how intertwined the care for human bodies is with the natural landscape.

Clear signs of real progress help people track changes without needing head-lines. In healthcare, these signs include fewer hospital visits and deaths over several years among groups receiving combined treatments compared to similar groups that don't. Insurance companies update their tables to show longer healthy lifespans.

Records show fewer bad outcomes as treatment methods improve. Progress is seen when insurance covers new treatments, doctors' groups update their care guidelines, and medical training includes these new methods.

Two Lives Unfolded

Maya

Maya's sprint times keep improving into her second century. She hasn't tran-scended physiology; her team trains inside rails. They have lowered risks by slight edits such as trimming lipids and clot threats, and tendon stiffness is treated to avoid rupture. They also measure heat dissipation to keep it within safe ranges.

There is a general registry that records adverse events in athletes across com-petitions, not just within a single laboratory. Maya competes in a declared enhancement division, whose rules and audits are publicly available. Part of her sponsorship funds a youth running program and a clinic that offers risk-low-ering care with long-horizon surveillance to families who could not otherwise afford it. When a cluster of arrhythmias appears among athletes trialing a new adrenergic modulator, the team pauses the protocol, publishes a postmortem, and treats the cohort with care. Winning remains a joy; safety remains the floor.

Evelyn

Evelyn makes instruments that sing. She will tell you that her hands are the best augmentation she has, and that what she has chosen to add are protections and minor adjustments. Risk-modifying therapies help prevent osteoarthritis from eroding her work productivity. A carefully calibrated neuromodulation protocol improves her fine-motor control in the hours she uses it; she declines a baseline change that would alter her tactile thresholds when she is not at the bench.

Once a year, she reviews her consent with her clinician, reads the outcomes and adverse events, and signs again with a pause built in for backing away if doubt arises. Her shop logs air quality and discloses finishes, and a QR code on the label of each instrument directs users to a "craft provenance" that lists the woods, methods, and care. She mentors trainees from under-resourced backgrounds, because the program that helped her requires those who benefit to keep the door open for others.

Failure Modes and How to Survive Them

In biological enhancement, problems occur when people focus on the status they get from a treatment instead of its safety and honesty. A clear example happened at a state sports event. The athlete, focusing more on the prestige than the safety protocols, suffered a severe tendon rupture. This incident highlighted the critical need for procedural repairs: implementing category frameworks for sport and performance, ensuring that provenance accompanies achievements, and conducting regular audits and enforcing proportionate consequences for deception.

Inequity drift is the second most relevant failure. The repair involves tying public money to equity metrics, publishing gaps, and reallocating budgets until the gaps close, again and again, without fatigue. Sensory overload and drift are common when augmentations transition too quickly from task-specific use to rest-state changes. The solution is to tie growth to boundary budgets and keep the bio-stewardship tithe visible enough that citizens can see reciprocity rather than rhetoric.

The last failure is legacy capture. Programs that succeed ossify; leaders who succeed often entrench themselves. The repair involves attaching time-to-live limits to control rights, utilizing sortition on oversight boards, and establishing sunset clauses and reauthorization windows that require written reasons and diffs future stewards can review.

Closing: Enhance Without Erasing

Choosing the biological path does not mean rejecting technology or judging other options; it simply means making an informed decision. It's about setting thoughtful defaults. "Enhance without erasing" encourages people to stay true to their bodies, support those who depend on them, and care for the living systems that make life meaningful.

Living this way gives your morning run a deeper meaning. It becomes part of a larger story that connects care for people and places, helping to ensure the future remains open for the next generation. We seem to be standing on a path of unlimited progress. How much further are we going to push the boundaries? Our new environment will be the universe itself.

Chapter 14

The Infinite Frontier

A Signal That Arrives Late and Still on Time

Mara starts her shift as the hydro loop dims, bringing a kind of artificial dusk. Gate-3, a ring at the edge of the Kuiper Belt, turns steadily, patient like a well-worn clock. It turns ice into air and water, builds structures from nickel, and sends ships out to places where sunlight barely reaches.

On Mara's console, a council docket from Earth arrives after the usual 16-hour delay. Its arrival is routine. A policy journal with changes, signatures, and a rollback plan appears just like any other message that follows the speed of light. Mara signs for it, reads the summary, and sets up a meeting for the day after tomorrow so everyone can decide what to accept, what to change, and what to send back with feedback.

A child's laughter rises in a soft feed from a family module two rings away, where morning is offset to match a caregiver's duty cycle. Somewhere behind her, a maintenance bot rattles a panel, then stops because the system it serves tells it to wait. A light across the bay settles from "busy" to "calm," and the air, which is work as much as it is element, tastes like the metal of living things. Mara feels a hint of relief as the docket's page count ends where it should.

Later in her shift, an alert flags a radiation spike upstream of a control circuit. The bot pauses again because the plan says so. The lights step down in a way the crew calls "graceful." The ship's decision assistant cues the drill they've practiced since training: the black-start sequence, the islanding of life-critical loads, and telemetry posted on a public page that anyone on the ring can access without credentials.

Mara's pulse quickens and then settles. The drill ends, the system returns, and no one claps. They sign the postmortem and patch the weak point. Someone will read the notes two years from now after another drill on another ring. This is how a ship becomes a city and a city becomes a school. Life moves on, along with constant learning.

What Changes When We Leave the Cradle

People often talk about space as if it's all about distance and light, but the reality is more practical and demanding. The two main limits that shape decisions are delay and scarcity. Delay manifests in the time it takes for messages to travel, for supplies to arrive, and for actions to yield results. The speed of light sets a rigid boundary, changing how people govern and engineer their world.

Scarcity isn't just about money. It also refers to having enough reliable power, measurable air, recoverable water, usable heat, and attention that doesn't run out. To draw a parallel, both Antarctic research stations and submarines offer terrestrial analogs where similar constraints govern life. In these isolated environments, delay and scarcity are equally dominant.

At the end of a long day, Mara sits down in the cramped dining area with her team for their evening meal. The rations are finite, each portion calculated to the gram to ensure sustainability until the next supply ship arrives, weeks away. Tonight, Mara rehearses a familiar ritual: She measures water into a personal cup with careful precision. As she does so, the steady pulse of the depot's systems in the background reminds her that every resource is precious. The conversation turns to whether they will need to stretch their current supplies even further, just in case. It's a small decision in the moment, but it depicts their lives in space: delay and scarcity pressing down like gentle but unyielding hands.

Bodies that evolved to live under 1 G and a sky that flickers day and night with regularity are asked to grow old under rotation and under a blackness that will not tell them what time it is. This has several evident consequences:

- Microgravity or its reduction under light spin reduces load.

- Bone bleeds minerals and grows brittle.

- Muscles lose cross-sectional area without the resistance of a floor.

- Fluids shift headward and push on eyes that evolved to sit beneath water and sky, and retinas complain.

Every one of those changes has been measured and mitigated in near-Earth testbeds, but when years replace weeks, we don't talk about mitigation any longer. We need to have maintenance procedures to rely on. Radiation tracks across DNA become present in aging people. Circadian disarray causes a lack of sleep and judgment in an imperceptible way.

Cognition also changes: first, because the cortex is suddenly more or less capable. Moreover, the problem it must solve has shifted. On Earth, many decisions can be made in a turn-taking loop. The delay we talked about breaks that rhythm. Missions spanning light minutes and light hours require asynchronous collaboration as a default. Software engineers refer to this style of tolerance for divergence as "CRDT" and "conflict-free," but there are ethical concerns.

Culture bends around delay in less technological, more urgently human ways. Long tours can turn crews into cities. Crafts become neighborhoods, and the crews that live on them for long periods to accomplish their missions inaugurate new ones. Meals together become part of the infrastructure, and holidays timed to orbits create collective memories.

The sound of a child in a module two rings over becomes part of the system. Intergenerational habitats reopens the question: What counts as "us" when the street is a rail and the horizon refuses to bend?

The answer? A reality we all build together.

Guiding Policy: Ships and Settlements Are Cities and Must Pass Audits

At this stage, life happens on spaceships as much as on Earth. These ships are home to entire communities who spend their whole lives on them. Thus, it is insufficient to name them simply as ships or crafts. They become cities. So, we treat every settlement and ship as a city with the same obligations that we learned on Earth.

Let's see how, through an example of an audit of a change in habitat policy regarding resource allocation. The process begins with a proposal reviewed by an appointed committee, which assesses the suggested change against current standards and future projections. The proposal includes a detailed impact study, followed by a trial phase where outcomes are monitored against predefined metrics.

Third-party auditors ensure the transparency of the data collection and analysis stages, and the results are made publicly available in a report. After the trial, the auditors collect feedback from inhabitants and evaluate it before a final review meeting. Only after reaching a consensus that the benefits outweigh the potential risks do they adopt the policy.

We can see that consent becomes a process again for baseline-altering inter-ventions such as sensory overlays, implants that change how a person perceives, merges that temporarily blend cognition, and memory-sharing that pushes ex-periences into a mind that did not live them. Any habitat or ship worth living in requires time-bounded consent that is renewable.

Engines, air handlers, water scrubbers, and medical bays are taught in school as biomes to ensure reliability. The habit of freezing expansions until failure paths are patched becomes part of how a ring survives.

Compute becomes part of the daily routine and plays new roles in familiar dynamics. Devices allow a child to talk to relatives on Earth, a maker to route a simulation, and a caregiver to run a diagnostic and book an appointment for the next morning. Queue fairness is public, so it isn't limited to those who can afford it.

The dashboards are adapted to new relevant metrics. A ring that wants to call itself a city must publish all of the following:

- oxygen levels, CO2 scrub capacity, water reclaim ratios, and nitrogen loop integrity

- contaminant loads

- the radiation people absorb at their workplace, including the variance

- where food grows and how resilient those systems are

- spare-part inventories for items that cannot be printed in their entirety

- thresholds for corrective action, with budgets moved when the numbers cross those thresholds

Bodies and Ships: A Biological Stack for Frontier Life

The body and the ship share the project of keeping a person fully themselves over the years, away from home. Old tests taught medicine that acute measures work.

Gravity and load anchor the beginning of a safe tour. Rotation that produces a proper fraction of a G is a gift. Where the ring can be big and the spin slow, joints sigh in relief and spines remember that upright is not a guess. Where the ring is small or the mission requires a wheel, the load must be substantial and heavy enough to stimulate the bone to remain and the muscle to grow.

People train with special programs that include resistance exercises with bars, rucks, and braces that are part of public gyms. Electrostimulation can help where compliance fails or where injury makes it necessary. Leisure is part of the training, and there is a schedule that reduces boredom; this is considered a medical intervention, as boredom can be just as detrimental as force.

Circadian timing matters too. In cities, light is shorthand for the calendar. On rings, light becomes a tool made by neighbors. Signaling sleep at the same hour every "day," recording quiet hours in a rule that is both medical and moral, and treating the screen like a candle rather than like a sun all helps. When duty shifts alter the sleep–wake cycle for a mission, the ship accommodates these changes, often in response to seasonal variations, rather than treating people as if they were clocks.

Temperature and pressure function like technical settings yet influence emotional states. A vessel that holds a constant temperature eliminates one fluctuating factor; a suit that effectively manages heat and clearly alerts the wearer to hazards eliminates another. Designers working on a ring habitat apply precise color and sound schemes to distinguish safe areas from dangerous ones, helping the brain reduce anxiety responses. Pressure remains within comfortable limits tailored to each crew, while discarding exaggerated claims of "low pressure" prevents respiratory harm.

Teams work on radiation risk. They track doses and display them with a report about the likelihood of facing a particular job on the subsequent tour. How many centimeters of regolith sit above a sleep ring is an argument at the assembly and ends in a budget line. Parameters that promise "biological radioprotection" are not led by this culture. There is room for pharmacology and for carefully supervised studies of repair pathways, and there is also the restraint to say, "No, this ex vivo protein remains where it belongs for now."

Microbiomes at the scale of the gut and a bioreactor become an indicator of health. Cities on rings plant gardens not only to absorb CO_2 but as a tool to resist depression.

Continuous monitoring exists because a closed system must watch itself if it wants to feel free. The monitors display data from a range of sources:

- Wearables track sleep, heart rate, and load.

- Cabin sensors count pollutants that stop at words like "trace" and then, when they rise, trigger decisions long before they cause fever.

- Readings do not "belong" to a person in the way a private letter might.

They belong to a network that says, "You are us, and we are responsible for keeping you well enough to choose work and love with your mind on those choices rather than on the murk in your air."

- Medical packs designed to treat decompression accidents, toxic exposures, and neurological emergencies sit beside the joyous modules and the quiet ones.

Perception, Interfaces, and Cognitive Safety

In a place where the world offers fewer cues than a person uses, and where information arrives late, we need a new design. Sometimes, more data is not always good enough.

Multi-spectrum sensing is both useful and tricky. It changes infrared light into signals the eye can see, turns vibrations into sensations felt through a glove, and shows radiation as moving shadows. These features help keep people safe. The system lowers these alerts when someone shows they know what they're doing and keeps performing well. It uses simple settings, limits how many alarms can go off, and accepts that some warnings might be missed to avoid overwhelming attention. The important thing is how the tool helps a person in real life.

Brain–computer interfaces (BCIs) are already used as medical tools on Earth, and they will be used for medicine here as well. They will also help in surgeries where speed and accuracy matter a lot. These devices must follow certain rules. People have to agree to use them again and again. The systems must know that they can sometimes fail when under stress, so they work in pairs, like a monitor working with a person or one person working with another, rather than expecting everything to work perfectly all the time. Using BCIs all the time can seem overconfident. Letting a person take off a tool and still be okay when they are not working shows a society that is learning to use technology to help life, not control it.

Managing mental effort is a community habit. It demands taking breaks to avoid staring at an overwhelming stream of information. Quiet times and spaces help people focus, and simple reminders and notifications make decisions easier, not harder. When someone is tired, they ask tools to write or post for them, not just to complete tasks, because the tools enhance clarity. Groups work with flexible timing instead of against it.

Governance Under Delay

Many issues emerge in this fascinating context because the landscape of reality has dramatically expanded. Among the most urgent issues are the need to make decisions without immediate communication and the innate human tendency to lean toward charismatic leadership when people lack quick feedback loops.

These challenges demand innovative solutions to support transparent decision-making and equitable governance. The new politics doesn't need new words to deal with delay; it only requires the will to stop playacting Earth where it has no power.

Ruling frameworks from Earth arrive at the pace physics allows. One ring reads, considers, and acts with reasons. The next ring writes its own policy journals with patterned sections for the crew:

- Why we think this is the problem.

- What we intend to do about it.

- What we commit to track and publish while we do it.

- What would change our minds.

- When we promise to revisit the choice.

- Under what measures will we roll it back if anyone feels embarrassed.

The rhythm repeats. Rhythm functions as the collective memory system that prevents traditions from becoming rigid and obsolete through mere generational change.

What will happen when a ring does not agree with the rules coming from the Earth? A ring that has diverged from an Earth council does not deny that divergence. When rights are limited to speed up a process, special protections for minority rights can move a case to the front of the line for a focused hearing. This is done under a set of rules that clearly define what can be reviewed and require written reasons for decisions.

Identity governance becomes both stricter and fairer because branch multiplication would otherwise become a civic exploit.

Ultimately, governance that tolerates delay becomes the default, rather than an awkward exception. Ships and settlements cannot hold a meeting with Earth that resolves a controversy in an afternoon and declare victory. They must replace "call and response" with policy journals that write "what we thought was going on," "what we chose to do," and "what would make us stop or change," and then they must be willing to read and reconcile weeks later when packets align.

Software proofs of this practice are the tools that engineers use to implement eventual consistency, which they refer to as conflict-free. Human proofs are the stories communities tell themselves about adopted policies that diverged and then merged with grace.

Stewardship and the "Public Face" of Closed Loops

Earth has learned to recognize planetary boundaries and to treat red–amber–green as a yes–no–maybe for budgets. Rings and ships turn the same habit inside out and use the dashboards as public documents to display resource usage. Can this ring sustain itself for three months if the next tanker is delayed? What can the ring do to eat better with what it has?

Spare-part circularity diagrams illustrate the frequency of a printed part being reused and whether bottlenecks exist. Every chart is accompanied by a method section, which serves to build trust and facilitate the reproduction of steps.

Identity, Continuity, and Parity

A ring that wants to avoid becoming an oligarchy needs to set clear and solid rules. Besides all the precautions we have explored previously, agents that wish to be recognized for specific purposes pass tests under supervision and are recognized within those narrow lines. The agent who says, "I am a doctor again" passes weeks of practice under observation before a patient's dignity is put at risk.

Psychological Scaffolding: Doing the Inner Work in a Vacuum

Long tours require a different kind of courage than a sprint. The courage in question is taking the slow route to success, as speed can break things that cannot be replaced.

Choosing ships is all about resilience and how well they work together. Teamwork matters. When a crisis hits, there's always someone who can handle disruptions, someone who can call for a break, and someone who keeps track of details without getting rattled. Work shifts are set in a way that's fair to both the union and management.

People living on the ships create new rituals and adapt others inherited from life on Earth. Holidays are anchored both to orbits and to Earth's seasons. Shared virtual spaces are used in various ways. Friends in Lagos, Kyoto, and the ring sit in a virtual kitchen and eat and talk, and the defaults respect attention by turning time budgets on and by putting a noticeable ridge in the path of binging. Moderators get paid and cared for because they absorb harm to prevent greater damage.

People keep journals and record audio stories, sometimes for their own enjoyment, and sometimes for children who have not yet been born. A person's ability to revoke a permission to use their memory in a training module is not a token. It is effective where technically possible. AI companions remind a person of a recipe and ask for permission before copying it into a shared book.

Cultural Continuity With Pride and Joy

There's a difference between lazy nostalgia and actually remembering things. Traditions and objects that last pick the latter. Digital archives come with clear information and context. You can scroll through museum exhibits on your phone, but they're also displayed in shared spaces because a kid learning geometry under a print of Hokusai, with a caption about waves, gains several important insights at once. Festivals are adjusted to fit the visible sky; they still carry the familiar smells of food from neighborhoods where buses had simple numbers instead of defined routes.

Children exchange messages as homework with kids in cities they have never been to, and their teacher is as much of a bore as a delay. People name rings, rooms, and small plants after elders who are alive, after rivers that no one has seen in person, and after a bot that saved a life. The names show thanks and the belief that distance doesn't change what's right or wrong.

Worked Example: Five Years of Gate-3

This example of Gate-3 is a synthesis of multiple analogous scenarios derived from experiments and modeled projections. This method helps identify patterns and evaluate the implications and challenges of long-term space habitation, lending it value as an explorative and educational narrative.

In the first year after "Gate-3 is a ship" becomes "Gate-3 is a place," the ring launches its public life-support dashboard. The team took longer than anyone wanted to get the method section right because it refuses to post numbers without teaching the population how to read them. On a standard screen, down a central corridor, oxygen production is represented by a blue line, carbon scrub capacity by a green rectangle, and the water reclaim ratio by a moving color wash.

Nitrogen loop integrity has become more of a performance than a technical issue. Radiation dose maps for workstations are quiet but frustrating images that affect work shifts rather than personal relationships. Spare-part inventories are usually shown in a simple table with a long, scrollable interface.

The election stack runs its first black-start and islanding drill. It fails. The ring admits that a switch was older than it claimed in the paperwork. Budgets move. Someone says "freeze" at a plenary, and no one argues. Then, the expansion stops, and it is a relief to watch the culture trust itself.

In the second year, a compute commons opens that spreads the sense that the ring belongs to those who can book cycles and leaves everyone else with scraps. Baseline cycles are now publicly guaranteed and display a list of projects with claims to bursts that apply for and receive windows for specific reasons. Anyone sitting at the console in the corridor can see the queue and the posted latencies. A woman comes in with a continuation, saying it belongs to her husband, but she doesn't get a confirmation immediately. Claims are tested over several months and have limits. Behavioral novelty tests are conducted during ship operations instead of in a lab.

Autobiographical recall with context grows less awkward as the agent finds pace. The ring says "no" to some claims until time has shown whether the agent can be trusted with more.

In the third year, a heat budget spike appears. It could be a check bouncing. The ring staff writes another public note that uses the word "pause." It is a habit by now. A planned import of compute racks and their cooling infrastructure would shatter the heat line for three months. The council promotes the upgrade that will make heat behave back to the top of the list. A week after the new loop comes online, the ring makes no change on the dashboard, because the point of the dashboard is not to celebrate.

The fourth year brings a policy change from Earth in a packet called "small." A citizens' assembly chosen by sortition reads the change, hears from experts, and then writes its own document called "adopt with modifications." A minority rights veto pauses a line that would have made it harder to appeal decisions about conflict under delay. The fast-track review returns a narrower version of the rule.

In the fifth year, the ring notices that three workstations are accumulating radiation at a faster rate than the others. The ring holds a radiological shelter drill that disrupts everyone's sleep. Later, they deliver the promised report. Duty rotations are adjusted, and additional shielding is installed. The compute commons distributes a small amount of money labeled as a "continuation compute tithe" to a belt-wide stewardship trust. This trust funds an archive for family stories that moves between rings with signatures attached.

A maintenance worker points out that the time to hear petitions is longer for their group compared to those who work at desks, and the concern is justified. A budget trigger is activated, prompting further changes. Three months later, the dashboard shows a meaningful improvement.

A Failure Log You Should Expect—And Survive

I have refused to talk about failure. Instead, I have talked about constant learning and adaptation. Ships and rings will fail in ways that are familiar and in ways that are specific to an extended tour under delay. The only honor in prediction is the honor of preparation.

Overload from multispectral overlays can cause people to break. If the training loop makes sensing a marathon rather than a mat class, people will fall. To avoid this, they can move even training under the discipline of duty cycles and design quiet defaults that a person can lift from without shame.

Identity drift without memory will break promises. People will forget what they were about. The ring will keep diaries mandatory and forgive mediocre writing in favor of writing at all. It will ask people, every decade, to write value diffs that say, "This is why I think differently," and to send them to those whose lives they have changed with those shifts. It will route obligations through branch IDs and continuity trusts.

Governance capture under delay will be easy. Charisma loves a vacuum almost as much as oxygen does. The ring will limit terms, rotate sortition, protect minority appeal, and trust sabbaticals to create better leaders than clinging dependence does. It will refuse to take attendance at every ritual and thus avoid decoupling from reality.

Resource capture by elites will reappear. Compute and power love winners unless winners disobey the rules. Some of them are queue fairness, sampled audits, and embodied energy and heat budgets stitched to expansion freezes. The ring will write those rules down.

Fantasy physics will tempt lonely hearts. Nothing is lonelier than a ring that believes it has been forgotten, and nothing attracts fever dreams like loneliness. The ring will teach why "instant links" are not a thing. It will write an essay titled "Entanglement is not a phone" and hand it out to visitors and teach it in school. It will keep delight for what is true, and it will keep hope away from superstition.

Ecological rebound will appear in miniature in places no one thought to look. A ring that becomes so efficient it forgets that efficiency invites overuse will watch its water and heat lines flinch. It will learn to say, "Enough for now" with grace and then train the teenager who wants to change the world to repeat the sentence.

Transcendence, Cautiously Named

Transcendence means life here is small and close. What makes a community strong is choosing to respect each other over time. Taking rest as seriously as checking equipment is a form of transcendence. When a council explains its decisions openly, or when a child learns both to bake bread and to use new tools, that's transcendence too. Even choosing not to take on a role you're not ready for is part of it. Here, transcendence is about moving from just being capable to truly caring for one another.

In psychology, the concept of self-actualization reflects a similar journey from potential to self-improvement and a focus on community. In sociology, such communal strength and care resonate with theories of social capital, which emphasize the importance of relationships, networks, and trust in maintaining societal health. Organizational studies also underscore the importance of transparent decision-making in effective governance and cooperation, thereby further grounding the relevance of practical transcendence across disciplines.[1]

Concluding Thoughts: The Same Work, Now With Stars

Rings are neighborhoods without fields to hide a mistake. The rails that made a town decent on Earth make a ring safe out here. The speed of light is a line we draw on our maps and do not cross.

If a person were to tap the glass of a corridor window at Gate-3 and wish someone they loved could be there to watch the cold, they could be forgiven for feeling a pang. The ring's political virtue is to feel that pang, and then to finish the drill and add a line to the journal; this way, the person they are wishing for can read it when the packet arrives and can be proud of the person who wrote it rather than merely moved by the desire.

Epilogue: The View From Tomorrow

Sarah Taylor stands in the observation gallery of the Luna Research Commons, watching Earth rise: a blue light with thin ribbons of aurora and nighttime grids. A slim slate sits on the sill beside her, a habit she's kept for decades. It displays the station's public stewardship dashboard: oxygen buffer, water reclaim ratio, scrub capacity, drill cadence, and the day's black-start rehearsal notes. The information is clear enough for a teenager to read. These numbers give her more reassurance than the view outside. To her, they show that beauty depends on order and function.

Two centuries have passed since she first annotated a paper about convergence in a student lab. Since then, she has lived in many places: working in labs that taught cells to grow, spending months in clinics where a pause in consent was as important as a scalpel, and spending years in virtual rooms that felt like any hall on Earth. She has lived as one person and, for a time, as more than one.

Nothing about her life feels like a switch flipped from "before" to "after."

Behind her, the Commons is alive with everyday activity. A pleasant murmur of conversation underlies the space. In one bay, a continuity agent reads with grandchildren who have only known their grandmother as a voice and a face on a screen, but who still recognize her by the familiar warmth of her laughter and the comforting scent of her lavender cushion. The official record of her recognition hangs on the wall. No one claims that passing these tests makes someone perfect; it simply does enough to keep families safe and let love continue.

In another bay, a maintenance team returns from a drill and posts their after-action report. The station managed to isolate its life-critical systems on time, though a slow valve readout caused 40 seconds of uncertainty. The report is honest about mistakes; budgets are adjusted before comments are made. The drill will happen again in three days, because repetition is a principle here, not just a routine.

Farther down the corridor, a council chamber opens. Branches decide among themselves who will represent them each season. Today's agenda focuses on expanding the compute commons for a school group that has outgrown its current space. The proposal includes round-the-clock clean power, a heat budget that stays within limits, and a clear plan for water use and return temperatures. No one here trusts a promise that hasn't already been tested.

On Sarah's slate, a small olive tile marks notes from her last visit to Gate-3, the Kuiper Belt ring that now calls itself a city instead of a ship. The habits she learned there have served her well.

People often say that "forever" feels like weightlessness. But to her, what keeps it grounded is a short list she repeats like a prayer:

- Consent renews on schedule and can be revoked without shame.

- Claims that invite consequences do not pass on charisma; they pass on preregistered tests under independent audit.

- Identity pluralizes without buying extra ballots; care follows presence and shows up in ledgers, not just in speeches.

- Access is disaggregated on dashboards, and budgets move when gaps persist.

- Energy and reliability are part of ethics; outages are rehearsed until surprise loses its sharpest teeth.

- Worlds, physical and virtual, publish their methods and their diffs.

Everyone on the station knows this list by heart. It represents the guiding principles of a society that chooses to renew itself intentionally, not just in response to loss.

Sarah remembers the funerals she attended before mortality became a choice. Some were riven by accident or injustice; some were ordinary and good. Suddenly, an alert sounds over the station's intercom, breaking her reverie. The system reports a minor atmospheric correction, a routine reminder that even at rest, their vigilance persists.

She acknowledges it before returning to her thoughts. She thinks of Maria, who chose an end in her own time in a sunlit room, and whose grandchildren later watched a different kind of goodbye when a beloved continuation's scope closed gently because that was the boundary she had set. She thinks of Aminatou, gone too soon in a city where budgets did not match ability, and of the national fund that bore her name afterward, with rules that made the following petition less a plea and more a process.

The station's subtle hum continues, keeping pace with her reflections. She thinks of Lena, who set plates for three versions of herself and learned to make a life with charters and ledgers, rituals of gratitude, and rules for memory that protected other people's privacy as carefully as her own. She thinks of the first station that published a black-start drill and was ridiculed by people who still believed that reliability could be promised instead of proven. She thinks of how quickly the laughter stopped when storms began to teach.

In the Commons garden, a meeting begins. Representatives from settlements in orbit and on rings join by presence or by proxy. The docket concerns something that sounds like science fiction: seeding a learning lattice to accompany a slow mission to Andromeda.

When the vote ends, the chamber's display lights up with the changes. The proposal passes with three amendments, including a sunset clause and a requirement for independent audits. There is no celebration, just the quiet satisfaction of a decision made for reasons anyone could read and question.

Between sessions, Sarah walks the long window where Earth, the Moon, and a rim of panels sweep together like clockwork. A colleague asks what she thinks about the young who call their lives distributed, their homes latticed, their loves branched. She smiles. "It feels like music," she says. "Polyphonic when it is good. Noise when it forgets the key."

Then her expression grows serious. "The way we keep the key is the same as it has always been. We don't let power get ahead of procedure. We don't mistake speed for care. We don't make the people who are tired carry the weight of our brilliance. We fail in daylight."

On her way back to the observation gallery, Sarah sees a display called "Ledger of living systems." Ongoing investments in a once-contested program have enabled the growth of resilient mangrove ecosystems that withstand severe storms, restored forests previously ravaged by floods, and created shared trails used by both pronghorn and local communities. Both failed and successful projects are listed openly, because being honest about past efforts helps build trust.

In the evening, she returns to the window. Earth moves as it always has. Somewhere on its night side, a family sits at a wooden table in a house that smells of soap and cardamom, opening a slate to read a message from a grandmother who is no longer embodied yet is still present in ways that have earned the word. Somewhere on its day side, a council in a midsize city holds a hearing about a data center that wants to expand and says yes with conditions.

Sarah thinks about the questions she and her cohort asked when they were young:

- Would intelligence outrun us?

- Would continuation hollow us out?

- Would an end to the old way of dying make meaning thin?

The answers that endured turned out to be about character:

- Whenever a claim invited consequences, they demanded verification.

- Whenever a promise leaned on infrastructure, they demanded drills.

- Whenever access drifted toward those who were already well placed, they demanded dashboards and budgets that corrected the drift.

- Whenever identity threatened to become a trick, they wrote registries and charters that made plurality legible and parity absolute.

- Whenever work pressed close to the line between helpful and harmful, they made consent a cadence rather than a ceremony.

- Whenever they were tired, they rested without apology.

- Whenever they were wrong, they wrote it down where someone else could learn.

She knows these habits don't make people perfect. They just help everyone act decently in public, and that's the kind of decency that endures.

There is space for things that remain unfinished. The next decade will bring surprises, some wonderful, some difficult. The systems that have brought us this far will help us through both.

Sarah returns to the window, seeing Earth as an old friend. Her generation didn't erase endings; they made them choices and built systems to keep those choices fair. They didn't promise perfection, just good process, and they kept that promise. They didn't move faster than light but learned to work better within their limits. If the universe feels more alive now, it's not because anything became magical, but because people learned to write rules and records, using both to leave more life behind than they took.

Continue the Conversation

The questions raised in *The Architecture of Forever* are only the beginning. Now you can continue the dialogue directly with the book itself.

Meet *The Architect,* an AI companion trained on the ideas and text of this work. Ask it about immortality, identity, digital continuation, or any theme that moved you. It doesn't just summarize; it reflects, interprets, and helps you explore how these ideas might unfold in your own life and time.

Start your conversation here:

Whether you want to clarify a concept, challenge an assumption, or imagine the world that comes next, The Architect is ready to talk.

Acknowledgements

First and foremost, I owe my deepest gratitude to my wife, Nicole, and to our dogs, Snoop and Ladybird. Their companionship, patience, and unwavering support gave me the time, space, and strength to bring this book to life.

I am profoundly thankful to my editors, Kalima Sahriie and Clara Dunne. Their sharp insights, tireless attention to detail, and encouragement helped shape my ideas into something far clearer and stronger than I could have achieved alone.

Writing this book has deepened my respect for all authors. What began as a lifelong passion for exploring the future has turned into an even greater appreciation for the immense effort, discipline, and dedication required to put words onto the page and carry them through to publication.

Finally, I am grateful to the broader community of thinkers, dreamers, and explorers who continue to imagine what lies ahead. This book stands on the shoulders of that ongoing conversation, and I am honored to have contributed a piece to it.

Glossary

Autobiographical recall with context: A verification method assessing whether a digital continuation can recall events with proper detail, emotional tone, and personal nuance.

Behavioral concordance: A test to determine whether a digital continuation behaves like the original person in new and unfamiliar situations.

Black-start drill: An infrastructure test simulating a total power failure to ensure that critical systems (e.g., hospitals, servers) can restart independently.

Branch/branch ID: A distinct version or copy of a person that has split off (e.g., a digital or hybrid self), assigned a unique ID in registries to track rights, roles, and responsibilities.

Civic parity: The principle of equal standing in political or community decision-making, regardless of digital enhancement, branching, or popularity.

Continuation compute tithe: A mandatory contribution (financial or in compute resource terms) to environmental restoration, levied on high-energy continuation services.

Continuity trust: A legal vehicle that manages shared identity assets (e.g., archived memories, obligations) among multiple branches or after death.

Convergence: The reinforcing interaction of AI, biotechnology, and neurotechnology to accelerate medical, cognitive, and governance innovations.

Digital continuation: A digitally maintained model of a person's cognition, behavior, and memories that may persist after biological death for limited functions.

Elective mortality: A future condition where death due to aging becomes optional rather than inevitable, allowing people to choose when and how to die.

Instantiation: Commonly used in programming and computer science. It refers to the process of creating a specific instance (or object) of a class or data structure. A class can be thought of as a blueprint or template that defines certain properties and behaviors, and instantiation is the act of using that blueprint to create an actual object in memory, which can then be used in the program.

Islanded operations: Operating a system independently of the larger grid or ecosystem during emergencies is vital for critical infrastructure.

Memory merge: A form of cognitive intimacy or integration where emotional or sensory recollections are shared between individuals or branches.

One identity, one civic vote: A governance principle designed to prevent amplified political influence by disallowing copies from voting or holding office multiple times.

Platform primacy: The growing condition where virtual spaces become the dominant or "primary" environments for life, work, and governance.

Preregistered test/journal: A trial or policy initiative with clearly defined goals, metrics, and stop conditions published before implementation.

Presence-weighted ledger: A household or civic accounting method that allocates rights and obligations based on demonstrated presence and participation.

Rails: Structured governance constraints (e.g., consent protocols, audits, time-to-live limits) that ensure innovation remains safe and ethical.

Reversibility classifications: A method of categorizing decisions by how easily they can be undone: two-way door (easy), hinged (medium), or one-way door (irreversible).

Sortition: Random selection (like a lottery) is used to assign public service or governance roles for greater equity and representation.

Stewardship dashboard: A public-facing performance tracker showing ecological, civic, and infrastructural metrics (e.g., water consumption, equity rates).

Sunset clause (time-to-live limit): A built-in expiration date for a law, role, or institution, requiring it to be revisited and actively renewed.

Value diff: A personal update log where individuals describe how their values or ethics have changed across time, and why.

Value drift: The gradual shift in a person's values over an extended period, which may undermine past commitments or ethical coherence.

Verification regime: A multilayered framework to verify claims from synthetic identities (like continuations) by testing identity, memory, behavior, and provenance.

Virtual solipsism: A risk in immersive digital life whereby users isolate themselves from shared social reality due to overpersonalized, frictionless environments.

Endnotes

Convergence Point

1. Khwaji et al., 2024

Earth's Immortals

1. Martínez & Bridge, 2012

2. Piraino et al., 1996

3. Seluanov et al., 2018

4. Escarcega et al., 2023

5. Jönsson et al., 2008

6. Martínez & Bridge, 2012

7. Wagner et al., 2011

8. Piraino et al., 1996

9. Lu et al., 2020

10. Buffenstein, 2008

11. Seluanov et al., 2018

12. Piovesan & Biondi, 2020

13. National Research Council, 1995

14. Huang et al., 2022

15. Lelarge et al., 2024

16. Lu et al., 2020

17. Yenari & Hemmen, 2011

18. Roark & Iffland, 2025

19. Roark & Iffland, 2025

20. Martínez & Bridge, 2012

21. Wagner et al., 2011

22. Jönsson et al., 2008

23. Escarcega, 2023

24. National Research Council, 1995

25. Piovesan & Biondi, 2020

Ecosystem Dynamics

1. Summerhayes et al., 2024

2. Richardson et al., 2023; Rockström et al., 2009; Steffen et al., 2015

3. Ministry of Primary Industries, n.d.

4. McLaughlin, 2025

5. Lakshmi, 2022

6. Convention on Biological Diversity, 2024

7. Dinerstein et al., 2020

8. World Health Organization, 2025

Living With Your Copies

1. Amunts et al., 2024; Klohs et al., 2025

2. Flood, 2017

3. Flood, 2017

4. Roberts, 2017; United Nations Economic and Social Council, 2024

5. Roberts, 2017

6. Roberts, 2017

7. Đurović & Willett, 2023

8. Ruhl et al., 2021

9. Ruhl et al., 2021

10. Roberts, 2017; Ruhl et al., 2021

11. Roberts, 2017

12. Gordon-Roth, 2019; Korfmacher, n.d.

13. *Personal Identity*, 2023

14. Ruhl et al., 2021

15. Ruhl et al., 2021

16. (Roberts, 2017)

17. Đurović & Willett, 2023

18. Lyreskog, 2021

19. Gordon-Roth, 2019; *Personal Identity*, 2023

20. Lyreskog, 2021

21. Breakwell, 2021

22. Đurović & Willett, 2023

23. Amunts et al., 2024; Klohs et al., 2025

24. Amunts et al., 2024; Klohs et al., 2025

25. Đurović & Willett, 2023

26. Roberts, 2017; Ruhl et al., 2021

27. Breakwell, 2021

28. Ruhl et al., 2021

29. United Nations Economic and Social Council, 2024

30. Ruhl et al., 2021

31. Đurović & Willett, 2023

32. Ruhl et al., 2021

33. Lyreskog, 2021

34. Đurović & Willett, 2023

35. Breakwell, 2021

36. Ruhl et al., 2021

37. Flood, 2017

38. Gordon-Roth, 2019; Korfmacher, n.d.; *Personal Identity*, 2023

39. Flood, 2017; United Nations Economic and Social Council, 2024

The Three Humanities

1. Franco et al., 2017; Jimenez et al., 2021

2. Partala, 2011

3. Brugada-Ramentol et al., 2022; Conrad et al., 2024

Population Plateau

1. International Energy Agency, 2024

2. Ostrom, 1990

3. United Nations Department of Economic and Social Affairs, n.d.

4. World Business Council for Sustainable Development, 2022

5. Cattaneo et al., 2022; Mahtta et al., 2022

6. Convention on Biological Diversity, 2024; International Energy Agency, n.d.

Economy After Scarcity

1. National Institute of Economic and Social Research, 2024

2. Franz, 2003

3. Boullier, 2009; Heitmayer, 2024

Governance for Immortals

1. Hoppe, 2010; Zuboff, 2022

2. Shiota, 2019

3. McDearmid, 2014

4. Bouttes, 2022; OECD Nuclear Energy Agency, 2015

5. Jacobs & Matthews, 2012

Love in the Long Game

1. Alexander, 2021

2. Lyreskog, 2021

The Burden of Infinite Choice

1. March, 1991; Sutton & Barto, 2018

2. Sharma, 2024

3. Scheffler, 2013

4. Kahneman & Tversky, 1979

5. Iyengar & Lepper, 2000; Schwartz, 2004

6. MacAskill, 2022

7. Bostrom, 2013

8. Scheffler, 2013

Virtual Worlds as Primary Reality

1. Csikszentmihalyi, 1990; Deci & Ryan, 2000; Partala, 2011

2. Firth et al., 2024

3. Csikszentmihalyi, 1990; Deci & Ryan, 2000; Partala 2011

4. Stormy, 2017

5. Firth et al., 2024; Maksatbekova & Argan, 2025

The Infinite Frontier

1. Sari, 2023

References

Alexander, S. (2021, December 2). Book review: Lifespan. *Astral Codex Ten*, https://www.astralcodexten.com/p/book-review-lifespan

Amunts, K., Axer, M., Banerjee, S., Bitsch, L., Bjaalie, J. G., Brauner, P., Brovelli, A., Calarco, N., Carrere, M., Caspers, S., Charvet, C. J., Cichon, S., Cools, R., Costantini, I., D'Angelo, E. U., De Bonis, G., Deco, G., DeFelipe, J., Destexhe, A., Dickscheid, T., ... Zaborszky, L. (2024). The coming decade of digital brain research: A vision for neuroscience at the intersection of technology and computing. *Imaging Neuroscience (Cambridge, Mass.)*, *2*, imag-2-00137. https://doi.org/10.1162/imag_a_00137

Arone, A., Ivaldi, T., Loganovsky, K., Palermo, S., Parra, E. , Flamini, W., & Marazziti, D. (2021). The burden of space exploration on the mental health of astronauts: A narrative review. *Clinical Neuropsychiatry*, *18*(5). https://doi.org/10.36131/cnfioritieditore20210502

Bartle, R. (2003). *Designing virtual worlds*. New Riders.

Benkler, Y. (2006). *The wealth of networks: How social production transforms markets and freedom*. Yale University Press.

Bin Rashid, A., & Kausik, A. K. (2024). AI revolutionizing industries world-wide: A comprehensive overview of its diverse applications. *Hybrid Advances*, *3*, 100277. https://doi.org/10.1016/j.hybadv.2024.100277

Bostrom, N. (2013). Existential risk prevention as global priority. *Global Policy*, *4*(1), 15–31. https://doi.org/10.1111/1758-5899.12002

Boullier, D. (2009). The attention industries: Moving beyond opinion and loyalty. *Réseaux*, *2*(154), 231-246. https://shs.cairn.info/journal-reseaux-2009-2-page-231?lang=en

Bouttes, J.-P. (2022). *Nuclear waste: A comprehensive approach (3)*. Fondation pour l'innovation politique (Fondapol). https://www.fondapol.org/en/study/nuclear-waste-a-comprehensive-approach-3/

Breakwell, G. M. (2021). Identity resilience: Its origins in identity processes and its role in coping with threat. *Contemporary Social Science*, *16*(5), 573–588. https://doi.org/10.1080/21582041.2021.1999488

Brugada-Ramentol, V., Bozorgzadeh, A., & Jalali, H. (2022). Enhance VR: A multisensory approach to cognitive training and monitoring. *Frontiers in Digital Health*, *4*, 916052. https://doi.org/10.3389/fdgth.2022.916052

Buffenstein, R. (2008). Negligible senescence in the longest living rodent, the naked mole-rat: Insights from a successfully aging species. *Journal of Comparative Physiology B*, *178*(4), 439-45. https://doi.org/10.1007/s00360-007-0237-5

Cattaneo, A, Adukia, A., Brown, D. L., Christiaensen, L., Evans, D. K., Haakenstad, A., McMenomy, T., Partridge, M., Vaz, S. S., & Weiss, D. J. (2022). Economic and social development along the urban–rural continuum: New opportunities to inform policy. *World Development*, *157*, 105941. https://doi.org/10.1016/j.worlddev.2022.105941

Clement, G. (2011). Fundamentals of space medicine (2nd ed.). Springer.

Conrad, M., Kablitz, D. & Schumann, S. (2024). Learning effectiveness of immersive virtual reality in education and training: A systematic review of findings. *Computers & Education: X Reality*, *4*, 100053. https://doi.org/10.1 016/j.cexr.2024.100053

Convention on Biological Diversity. (2024, October 1). Kunming–Montreal Global Biodiversity Framework.Montreal: CBD Secretariat. https://www.cbd .int/gbf

Craig, M. A., Rucker, J. M., & Richeson, J. A. (2018). Racial and political dynamics of an approaching 'majority–minority' United States. *The Annals of the American Academy of Political and Social Science*, *677*(1), 204–214. https ://doi.org/10.1177/0002716218766269

Craig, R. K., Garmestani, A. S., Allen, C. R., Arnold, C. A. (T.), Birgé, H., DeCaro, D. A., Fremier, A. K., Gosnell, H., & Schlager, E. (2017). Balancing stability and flexibility in adaptive governance: An analysis of tools available in U.S. environmental law. *Ecology and Society*, *22*(2), 3. https://doi.org/10.575 1/ES-08983-220203

Csikszentmihalyi, M. (1990). *Flow: The psychology of optimal experience*. Harper & Row.

Cucinotta, F. A., Kim, M.-H. Y., & Chappell, L. J. (2013). *Space radiation cancer risk projections and uncertainties—2012. NASA Technical Paper TP-2013-217375*. NASA. https://three.jsc.nasa.gov/articles/TP_2013_Canc erRisk.pdf

Danforth, M. M. (2025). *Christian morality: Our response to God's love. Teacher guide.* Saint Mary's Press. https://www.smp.org/resourcecenter/reso urce/14227/

Damaševičius, R., & Sidekerskienė, T. (2024). Virtual worlds for learning in metaverse: A narrative review. *Sustainability, 16*(5), 2032. https://doi.org/10. 3390/su16052032

Davenport, T. H., & Beck, J. C. (2001). *The attention economy: Understanding the new currency of business.* Harvard Business School Press.

Deci, E. L., & Ryan, M. (2000). The 'what' and 'why' of goal pursuits: Human needs and the self-determination of behavior. *Psychological Inquiry, 11*(4), 227–268. https://doi.org/10.1207/S15327965PLI1104_01

Dinerstein, E., Joshi, A. R., Vynne, C., Lee, A. T. L., Pharand-Deschênes, F., França, M., Fernando, S., Birch, T., Burkart, K., Asner, G. P., & Olson, D. (2020). A 'global safety net' to reverse biodiversity loss and stabilize Earth's climate. *Science Advances, 6*(36), eabb2824. https://doi.org/10.1126/sciadv.a bb2824

Đurović, M., & Willett, C. (2023). A legal framework for using smart contracts in consumer contracts: Machines as servants, not masters. *Modern Law Review, 86*(6), 1390–1421. https://doi.org/10.1111/1468-2230.12817

Escarcega, R. D., Patil, A. A., Meyer, M. D., Moruno-Manchon, J. F., Silvagnoli, A. D., McCullough, L. D., & Tsvetkov, A. S. (2023). The Tardigrade damage suppressor protein Dsup promotes DNA damage in neurons. *Molecular Cell Neurosciences, 125*, 103826. https://doi.org/10.1016/j.mcn.2023.10 3826

Farisco, M., Evers, K., & Changeux, J.-P. (2024). Is artificial consciousness achievable? Lessons from the human brain. *Neural Networks, 180*, 106714. https://doi.org/10.1016/j.neunet.2024.106714

Feng, T., Jin, C., Liu, J., Zhu, K., Tu, H., Cheng, Z., Lin, G., & You, J. (2024). How far are we from AGI: Are LLMs all we need? arXiv. https://arxiv.org/abs/2405.10313

Firth, J., Torous, J., López-Gil, J. F., Linardon, J., Milton, A., Lambert, J., Smith, L., Jarić, I., Fabian, H., Vancampfort, D., Onyeaka, H., Schuch, F. B., & Firth, J. A. (2024). "From 'online brains' to 'online lives': Understanding the individualized impacts of internet use across psychological, cognitive and social dimensions. *World Psychiatry, 23*(2), 176–190. https://doi.org/10.1002/wps.21188

Fischer, J. M. (2022). Death, immortality, and meaning in life: Precis and further reflections. *The Journal of Ethics, 26*, 341–359. https://doi.org/10.1007/s10892-022-09392-8

Fishkin, J. S. (2018). *Democracy when the people are thinking: Revitalizing our politics through public deliberation.* Oxford University Press.

Flood, C. M. (2017). *The body satyrical: Satire and the corpus mysticum during crises of fragmentation in Late Medieval and Early Modern France* [Doctoral thesis, UCLA]. UCLA Electronic Theses and Dissertations. https://escholarship.org/uc/item/86t3c90c

Franco, L. S., Shanahan, D. F., Fuller, R. A. (2017). A review of the benefits of nature experiences: More than meets the eye. *International Journal of Environmental Research and Public Health, 14*(8), 864. https://doi.org/10.3390/ijerph14080864

Frantz, R. (2003). Herbert Simon. Artificial intelligence as a framework for understanding intuition. *Journal of Economic Psychology*, *24*(2), 265–277. https://doi.org/10.1016/S0167-4870(02)00207-6

Gilster, P. (2020, August 21). *Homo stellaris: Space and human transformation*. Centauri Dreams. https://www.centauri-dreams.org/2020/08/21/hom o-stellaris-space-and-human-transformation/

Goldhaber, M. H. (1997). The attention economy and the net. *First Monday*, *2*(4). https://doi.org/10.5210/fm.v2i4.519

Gordon-Roth, J. (2019, February 11). *Locke on personal identity*. Stanford Encyclopedia of Philosophy. https://plato.stanford.edu/entries/locke-person al-identity/

Gottman, J. M., & Gottman, J. S. (2015). *10 principles for doing effective couples therapy*. W. W. Norton.

Gottman, J. M., & Levenson, R. W. (1992). Marital processes predictive of later dissolution: Behavior, physiology, and health. *Journal of Personality and Social Psychology*, *63*(2), 221–233. https://doi.org/10.1037/0022-3514.63.2. 221

Haimila, R., and Muraja, E. (2021). A sense of continuity in mortality? Exploring science-oriented Finns' views on afterdeath. *OMEGA—Journal of Death and Dying*, *85*(1), 154–178. https://doi.org/10.1177/0030222821103 8820

Harwood, S., & Eaves, S. (2020). Conceptualising technology, its development and future: The six genres of technology. *Technological Forecasting and Social Change*, 161, 120174. https://doi.org/10.1016/j.techfore.2020.120174

Hassabis, D., Kumaran, D., Summerfield, C., & Botvinick, M. (2017). Neuroscience-inspired artificial intelligence. *Neuron, 95*(2), 245–58. https://doi.org/10.1016/j.neuron.2017.06.011

Heitmayer, M. (2024). The second wave of attention economics: Attention as a universal symbolic currency on social media and beyond. *Interacting with Computers, 37*(1), 18–29. https://doi.org/10.1093/iwc/iwae035

Herrfahrdt-Pähle, E., Schlüter, M., Olsson, P., Folke, C., Gelcich, S., & Pahl-Wostl, C. (2020). Sustainability transformations: Socio-political shocks as opportunities for governance transitions. *Global Environmental Change, 63*, 102097. https://doi.org/10.1016/j.gloenvcha.2020.102097

Hoppe, R. (2010). *The governance of problems: Puzzling, powering and participation*. Bristol University Press.

Huang, W., Hickson, L. J., Eirin, A., Kirkland, J. L., & Lerman, L. O. (2022). Cellular senescence: The good, the bad and the unknown. *Nature Reviews Nephrology, 18*, 611–627. https://doi.org/10.1038/s41581-022-00601-z

Human enhancement: The scientific and ethical dimensions of striving for perfection. (2016, July 26). Pew Research Center. https://www.pewresearch.org/science/2016/07/26/human-enhancement-the-scientific-and-ethical-dimensions-of-striving-for-perfection-2/

Ienca, M., & Andorno, R. (2017). Towards new human rights in the age of neuroscience and neurotechnology. *Life Sciences, Society and Policy, 13*, 5. https://doi.org/10.1186/s40504-017-0050-1

Intergovernmental Panel on Climate Change (IPCC). (2023). *AR6 synthesis report: Climate change 2023*. https://www.ipcc.ch/report/sixth-assessment-report-cycle/

Intergovernmental Science-Policy Platform on Biodiversity and Ecosystem Services (IPBES). (2019). Global assessment report on biodiversity and ecosystem services. https://ipbes.net/global-assessment

International Energy Agency (IEA). (n.d.). Data centres and data transmission networks. https://www.iea.org/energy-system/buildings/data-centres-and-data-transmission-networks

Internet Engineering Task Force (IETF). 2007, April. Delay-tolerant networking architecture.RFC 4838. https://www.rfc-editor.org/rfc/rfc4838

Iyengar, S. S., & Lepper, M. R. (2000). When choice is demotivating: Can one desire too much of a good thing? *Journal of Personality and Social Psychology*, *79*(6), 995–1006. https://psycnet.apa.org/doi/10.1037/0022-3514.79.6.995

Jacobs, A. M., & Matthews, J. C. (2012). Why do citizens discount the future? Public opinion and the timing of policy consequences. *British Journal of Political Science*, *42*(4), 903–935. https://doi.org/10.1017/S000712341200117

Jimenez, M. P., DeVille, N. V., Elliott, E. G., Schiff, J. E., Wilt, G. E., Hart, J. E., & James, P. (2021). Associations between nature exposure and health: A review of the evidence. *International Journal of Environmental Research and Public Health*, *18*(9), 4790. https://doi.org/10.3390/ijerph18094790

Jones, H. (2008). *The dynamic impact of EVA on lunar outpost life support.* SAE Technical Paper 2008-01-2017. https://doi.org/10.4271/2008-01-2017

Jönsson, K. I., Rabbow, E., Schill, R. O., Harms-Ringdahl, M., & Rettberg, P. (2008). Tardigrades survive exposure to space in low Earth orbit. *Current Biology, 18*(17), R729–R731. https://doi.org/10.1016/j.cub.2008.06.048

Kahneman, D., & Tversky, A. (1979). Prospect theory: An analysis of decision under risk. *Econometrica, 47*(2), 263–291. https://doi.org/10.2307/191 4185

Khwaji, A. H., Daghriri, M. A., & Awaji, F. A. M. (2024). The impact of laboratory automation and AI on healthcare delivery: A systematic review. *Journal of Health Sciences and Nursing, 10*(2), 15–20. https://doi.org/10.535 55/hsn.v10i2.6184.

Klohs, J., Chen, W. C., & Araki, R. (2025). Advanced preclinical functional magnetic resonance imaging of the brain. *npj Imaging, 3*, 27. https://doi.org/ 10.1038/s44303-025-00085-z

Korfmacher, C. (n.d.). *Personal identity*. Internet Encyclopedia of Philosophy. Accessed August 8, 2025. https://iep.utm.edu/person-i/

Kurzweil, R. (2005). *The singularity is near*. Viking.

Lakshmi, R.B. (2022, August 9). *3 lessons from Indian tribes on ecosystem conservation*. Earth.org. https://earth.org/indian-tribes-ecosystem-conservatio n/

Landemore, H. (2020). *Open democracy: Reinventing popular rule for the twenty-first century*. Princeton University Press.

Lee, R. (2003). The demographic transition: Three centuries of fundamental change. *Journal of Economic Perspectives, 17*(4), 167–190. https://doi.org/10. 1257/089533003772034943

Lelarge, V., Capelle, R., Oger, F., Mathieu, T., & Le Calvé, B. (2024). Senolytics: from pharmacological inhibitors to immunotherapies, a promising future for patients' treatment. *npj Aging*, *10*(12). https://doi.org/10.1038/s41514-024-00138-4

Liu, C., Meng, S., Zheng, W., & Zhou, Z. (2025). Research on the impact of immersive virtual reality classroom on student experience and concentration. *Virtual Reality*, *29*, 82. https://doi.org/10.1007/s10055-025-01153-w

Lockhart, C., Lee, C. H. J., Sibley, C. G., & Osborne, D. (2022). The sanctity of life: The role of purity in attitudes towards abortion and euthanasia. *International Journal of Psychology*, *57*(5), 613–27. https://doi.org/10.1002/ijop.12877

Lu, Y., Brommer, B., Tian, X., Krishnan, A., Meer, M., Wang, C., Vera, D. L., Zeng, Q., Yu, D., Bonkowski, M. S., Yang, J. H., Zhou, S., Hoffmann, E. M., Karg, M. M., Schultz, M. B., Kane, A. E., Davidsohn, N., Korobkina, E., Chwalek, K., Rajman, L. A., ... Sinclair, D. A. (2020). Reprogramming to recover youthful epigenetic information and restore vision. *Nature*, *588*(7836): 124–129. https://doi.org/10.1038/s41586-020-2975-4

Lyreskog, D. M. (2021). Withering minds: Towards a unified embodied mind theory of personal identity for understanding dementia. *Journal of Medical Ethics*, *49*, 699-706. https://doi.org/10.1136/medethics-2021-107381

MacAskill, W. (2022). *What we owe the future*. Basic Books.

Mahtta, R., Fragkias, M., Güneralp, B., Mahendra, A., Reba, R., Wentz, E. A., & Seto, K. C. (2022). Urban land expansion: The role of population and economic growth for 300+ cities. *Urban Sustainability*, *2*, 5. https://doi.org/10.1038/s42949-022-00048-y

Maksatbekova, A., & Argan, M. (2025). Virtual reality's dual edge: Navigating mental health benefits and addiction risks across timeframes. *Current Psychology*, 44, 6469–6480. https://doi.org/10.1007/s12144-025-07623-3

March, J. G. (1991). Exploration and exploitation in organizational learning."*Organization Science*, 2(1), 71–87.

Marsh, E., Vallejos, E. P., & Spence, A. (2022). The digital workplace and its dark side: An integrative review. *Computers in Human Behavior*, 128, 107118. https://doi.org/10.1016/j.chb.2021.107118

Martínez, D. E., & Bridge, D. (2012). "Hydra, the everlasting embryo, confronts aging. *The International Journal of Developmental Biology*, 56(6–8), 479–487. http://dx.doi.org/10.1387/ijdb.113461dm

McDearmid, R. (2014). *Time out and time off: A systematic review of the benefits of sabbatical* [MBA project]. University of Prince Edward Island. https://islandscholar.ca/sites/default/files/2024-12/ir_9546_pdf.pdf

McLaughlin, J. (2025, December 5). *California's cap-and-trade program generated $33 billion since 2006.* The Epoch Times. https://www.theepochtimes.com/us/californias-cap-and-trade-program-generated-33-billion-since-2006-post-5855421

Ministry of Primary Industries. (n.d.). *Quota management system.* https://www.mpi.govt.nz/fishing-aquaculture/fisheries-management/quota-management-system/

National Institute of Economic and Social Research. (2024, January 8). Exploring post-scarcity. *National Institute of Economic and Social Research.* https://niesr.ac.uk/blog/exploring-post-scarcity

National Research Council. (1995). *Natural climate variability on decade-to-century time scales.* The National Academies Press. https://nap.nati onalacademies.org/read/5142/chapter/1

Newman, S. J. (2024). *Supercentenarian and remarkable age records exhibit patterns indicative of clerical errors and pension fraud.* bioRxiv. https://doi.or g/10.1101/704080

Nielsen, M. A., & Chuang, I. L. (2010). *Quantum computation and quantum information* (10th anniversary ed.). Cambridge University Press.

OECD Nuclear Energy Agency (NEA). (2015). *Stakeholder involvement in decision making: A short guide to issues, approaches and resources.* https://www.oecd-nea.org/jcms/pl_14894/stakeholder-involvement-i n-decision-making-a-short-guide-to-issues-approaches-and-resources

O'Neill, O. (2002). *A question of trust: The BBC Reith Lectures 2002.* Cambridge University Press.

Ostrom, E. (1990). *Governing the commons.* Cambridge University Press.

Parfit, D. (1984). *Reasons and persons.* Oxford University Press.

Partala, T. (2011). Psychological needs and virtual worlds: Case Second Life. *International Journal of Human-Computer Studies, 69*(12), 787–800. https://doi.org/10.1016/j.ijhcs.2011.07.004

Pasupuleti, M. K. (2025). The quantum code of singularity: AI, life sciences, and the future of human immortality. In M. K. Pasupuleti (Ed.), *The quantum singularity: AI, quantum biology, and the future of human longevity* (pp. 120–30). https://doi.org/10.62311/nesx/68168

Pavez Loriè, E., Baatout, S., Choukér, A., Buchheim, J. I., Baselet, B., Dello Russo, C., Wotring, V., Monici, M., Morbidelli, L., Gagliardi, D., Stingl, J. C., Surdo, L., & Yip, V. L. M. (2021). The future of personalized medicine in space: from observations to countermeasures. *Frontiers in Bioengineering and Biotechnology*, *9*, 739747. https://doi.org/10.3389/fbioe.2021.739747

Pelligra, V., & Sacco, P. L. (2023). Searching for meaning in a post-scarcity society: Implications for creativity and job design. *Frontiers in Psychology*, *14*, 1198424. https://doi.org/10.3389/fpsyg.2023.1198424

Personal identity. (2023, June 2). Stanford Encyclopedia of Philosophy. https://plato.stanford.edu/entries/identity-personal/

Pike, G. (2020, November 26). *Euthanasia and assisted suicide—When choice is an illusion and informed consent fails*. BIOS Centre. https://bioscentre.org/articles/euthanasia-and-assisted-suicide-when-choice-is-an-illusion-and-informed-consent-fails/

Piovesan, G., & Biondi, F. 2020. On tree longevity. *New Phytologist*, *229*(3), 1318–1337. https://doi.org/10.1111/nph.17148

Piraino, S., Boero, F., Aeschbach, B., & Schmid, V. (1996). Reversing the life cycle: Medusae transforming into polyps and cell transdifferentiation in *Turritopsis nutricula* (Cnidaria, Hydrozoa). *The Biological Bulletin*, 190(3), 302–312. https://doi.org/10.2307/1543022

Preston, S. H., Heuveline, P., & Guillot, M. (2001). *Demography: measuring and modeling population processes*. Blackwell.

Raworth, K. (2017). *Doughnut economics: Seven ways to think like a 21st-century economist*. Random House.

Richardson, K., Steffen, W., Lucht, W., Bendtsen, J., Cornell, S. E., Donges, J. F., Drüke, M., Fetzer, I., Bala, G., von Bloh, W., Feulner, G., Fiedler, S., Gerten, D., Gleeson, T., Hofmann, M., Huiskamp, W., Kummu, M., Mohan, C., Nogués-Bravo, D., Petri, S., ... Rockström, J. (2023). Earth beyond six of nine planetary boundaries. *Science Advances*, *9*(37), eadh2458. https://doi.org/10.1126/sciadv.adh2458

Roark, K. M., & Iffland, P. H., II. (2025). Rapamycin for longevity: The pros, the cons, and future perspectives. *Frontiers in Aging*, *6*. https://doi.org/10.3389/fragi.2025.1628187

Roberts, M. (2017, November 12). Value, class and capital. *The Next Recession*. https://thenextrecession.wordpress.com/2017/11/12/value-class-and-capital/

Rockström, J., Steffen, W., Noone, K., Persson, A., Chapin, F. S., 3rd, Lambin, E. F., Lenton, T. M., Scheffer, M., Folke, C., Schellnhuber, H. J., Nykvist, B., de Wit, C. A., Hughes, T., van der Leeuw, S., Rodhe, H., Sörlin, S., Snyder, P. K., Costanza, R., Svedin, U., Falkenmark, M., ... Foley, J. A. (2009). A safe operating space for humanity. *Nature*, *461*(7263): 472–475. https://doi.org/10.1038/461472a

Ruhl, J. B., Cosens, B., & Soininen, N. (2021). "Resilience of legal systems: Toward adaptive governance. In M. Ungar (Ed.), *Multisystemic resilience: Adaptation and transformation in contexts of change* (pp. 509–529). Oxford Academic. https://doi.org/10.1093/oso/9780190095888.003.0027

Sari, R. (2023). Enhancing corporate governance through effective oversight and accountability. *Advances: Jurnal Ekonomi & Bisnis*, *1*(6), 344–356. https://doi.org/10.60079/ajeb.v1i6.291

Scheffler, S. (2013). *Death and the afterlife*. Oxford University Press.

Schwartz, B. (2004). *The paradox of choice: Why more is less.* Harper Perennial.

Seluanov, A., Gladyshev, V.N., Vijg, J., Gorbunova, V. (2018). Mechanisms of cancer resistance in long-lived mammals. Nature Review Cancer, 18(7), 433-441. https://doi.org/10.1038/s41568-018-0004-9

Shapiro, M., Preguiça, N., Baquero, C., & Zawirski, M. (2011). Conflict-free replicated data types. *Proceedings of the 13th International Symposium on Stabilization, Safety, and Security of Distributed Systems,* 386–400.

Sharma, S. (2024, July 7). *The paradox of choice.* Medium. https://medium.com/design-bootcamp/the-paradox-of-choice-why-more-options-can-lead-to-less-happiness-05e8d6f81b78

Sherman, W. R., & Craig, A. B. 2019. *Understanding virtual reality* (2nd ed.). Morgan Kaufmann.

Shiota, J. (2019). The rise and fall of the Icelandic Constitutional Reform Movement:The interaction between social movements and party politics. *Journal of International Cooperation Studies, 27*(1), 157–174. https://www.research.kobe-u.ac.jp/gsics-publication/jics/27-1/shiota_27-1.pdf

Simon, H. A. (1971). Designing organizations for an information-rich world. In M. Greenberger (Ed.), *Computers, communications, and the public interest* (pp. 37–72). Johns Hopkins Press.

Slater, M., & Sanchez-Vives, M. V. (2016). Enhancing our lives with immersive virtual reality. *Frontiers in Robotics and AI, 3,* 74. https://doi.org/10.3389/frobt.2016.00074

Solez, K., Bernier, A., Crichton, J., Graves, H., Kuttikat, P., Lockwood, R., Marovitz, W. F., Monroe, D., Pallen, M., Pandya, S., Pearce, D., Saleh, A., Sandhu, N., Sergi, C., Tuszynski, J., Waugh, E., White, J., Woodside, M., Wyndham, R., Zaiane, O., ... Zakus, D. (2013). "Bridging the gap between the technological singularity and medicine: Highlighting a course on technology and the future of medicine. *Global Journal of Health Science, 5*(6): 112–25. https://doi.org/10.5539/gjhs.v5n6p112

Steffen, W., Richardson, K., Rockström, J., Cornell, S. E., Fetzer, I., Bennett, E. M., Biggs, R., Carpenter, S. R., de Vries, W., de Wit, C. A., Folke, C., Gerten, D., Heinke, J., Mace, G. M., Persson, L. M., Ramanathan, V., Reyers, B., & Sörlin, S. (2015). Planetary boundaries: Guiding human development on a changing planet. *Science, 347*(6223), 1259855. https://doi.org/10.1126/scien ce.1259855

Stormy. (2017, September 17). *Eyes on the street*. Medium. https://medium .com/i-cities/eyes-on-the-street-ab12b39b960b

Summerhayes, C. P., Zalasiewicz, J., Head, M. J., Syvitski, J., Barnosky, A. D., Cearreta, A.,Fiałkiewicz-Kozieł, B., Grinevald, J., Leinfelder, R., McCarthy, F. M. G., McNeill, J. R., Saito, Y., Wagreich, M., Waters, C. N., Williams, M. & Zinke, J. (2024). The future extent of the Anthropocene epoch: A synthesis. *Global and Planetary Change, 242*, 104568. https://doi.org/10.1016/j.glopla cha.2024.104568

Sutton, R. S., & Barto, A. G. (2018). *Reinforcement learning: An introduction* (2nd ed.). MIT Press.

Tan, K. H. (2025). *The irreducible singularity of consciousness: A quantum-temporal analysis of identity, cloning, and digital immortality* [Thesis]. University of Suffolkhttps://doi.org/10.13140/RG.2.2.17755.68647

Tjin, T. P. (n.d.). *Principles of democracy: Majority rule, minority rights*. New Naratif. https://newnaratif.com/majority-rule-minority-rights/

Törnberg, P. (2025). Social media imaginaries and the city: How the attention economy is reshaping urban built environments. *Social Media + Society*, *11*(1). https://doi.org/10.1177/20563051251323389

United Nations Department of Economic and Social Affairs (UN DESA), Population Division. (n.d.). *World population prospects 2024*. https://population.un.org/wpp/

United Nations Economic and Social Council. (2024, December 27). *Review and appraisal of the implementation of the Beijing Declaration and Platform for Action and the outcomes of the twenty-third special session of the General Assembly : Report of the Secretary-General*. United Nations. https://docs.un.org/en/E/CN.6/2025/3

United Nations Environment Programme (UNEP). (2023, September 30). *Global framework on chemicals - For a planet free of harm from chemicals and waste*. https://www.unep.org

Uy, A. (2025, July 4). *The worm that can regenerate its entire body from just a tiny fragment*. Discover Wild Science. https://discoverwildscience.com/the-worm-that-can-regenerate-its-entire-body-from-just-a-tiny-fragment-1-285373/

Wagner, D. E., Wang, I. E., & Reddien, P. W. (2011). *Clonogenic Neoblasts Are Pluripotent Adult Stem Cells That Underlie Planarian Regeneration."Science*, *332*(6031), 811–816. https://doi.org/10.1126/science.1203983

World Business Council for Sustainable Development (WBCSD). 2022, May 11. *Circular transition indicators v3.0*. https://www.wbcsd.org/resources/circular-transition-indicators-v3-0/

World Health Organization. (2025, February 18). *Biodiversity*. https://ww w.who.int/news-room/fact-sheets/detail/biodiversity

Yenari, M. A., & Hemmen, T. M. (2010). Therapeutic hypothermia for brain ischemia: Where have we come and where do we go? *Stroke, 41*(10 Suppl. 1), 72–74. https://doi.org/10.1161/STROKEAHA.110.595371

Youvan, D. C. (2025). The propagation of consciousness: Why AI may succeed where humans cannot. https://doi.org/10.13140/RG.2.2.26056.84483

Zhang, C., Lakens, D., & IJsselsteijn, W. A. 2021. Theory integration for lifestyle behavior change in the digital age: An adaptive decision-making framework. *Journal of Medical Internet Research, 23*(4), e17127. https://doi.org/1 0.2196/17127

Zou, Y. (2024). Genetic enhancement from the perspective of transhumanism: Exploring a new paradigm of transhuman evolution. *Medicine, Health Care and Philosophy, 27*(4), 529–44. https://doi.org/10.1007/s11019-024-10 224-9

Zuboff, S. (2022). Surveillance capitalism or democracy? The death match of institutional orders and the politics of knowledge in our information civilization. *Organization Theory, 3*(4), 1–35. https://doi.org/10.1177/2631787722 1129290

Zubrin, R., & Wagner, R. (2011). *The case for Mars* (Updated ed.). Free Press.

www.ingramcontent.com/pod-product-compliance
Lightning Source LLC
Chambersburg PA
CBHW032051020426
42335CB00011B/279

* 9 7 8 1 9 6 9 9 8 9 0 3 2 *